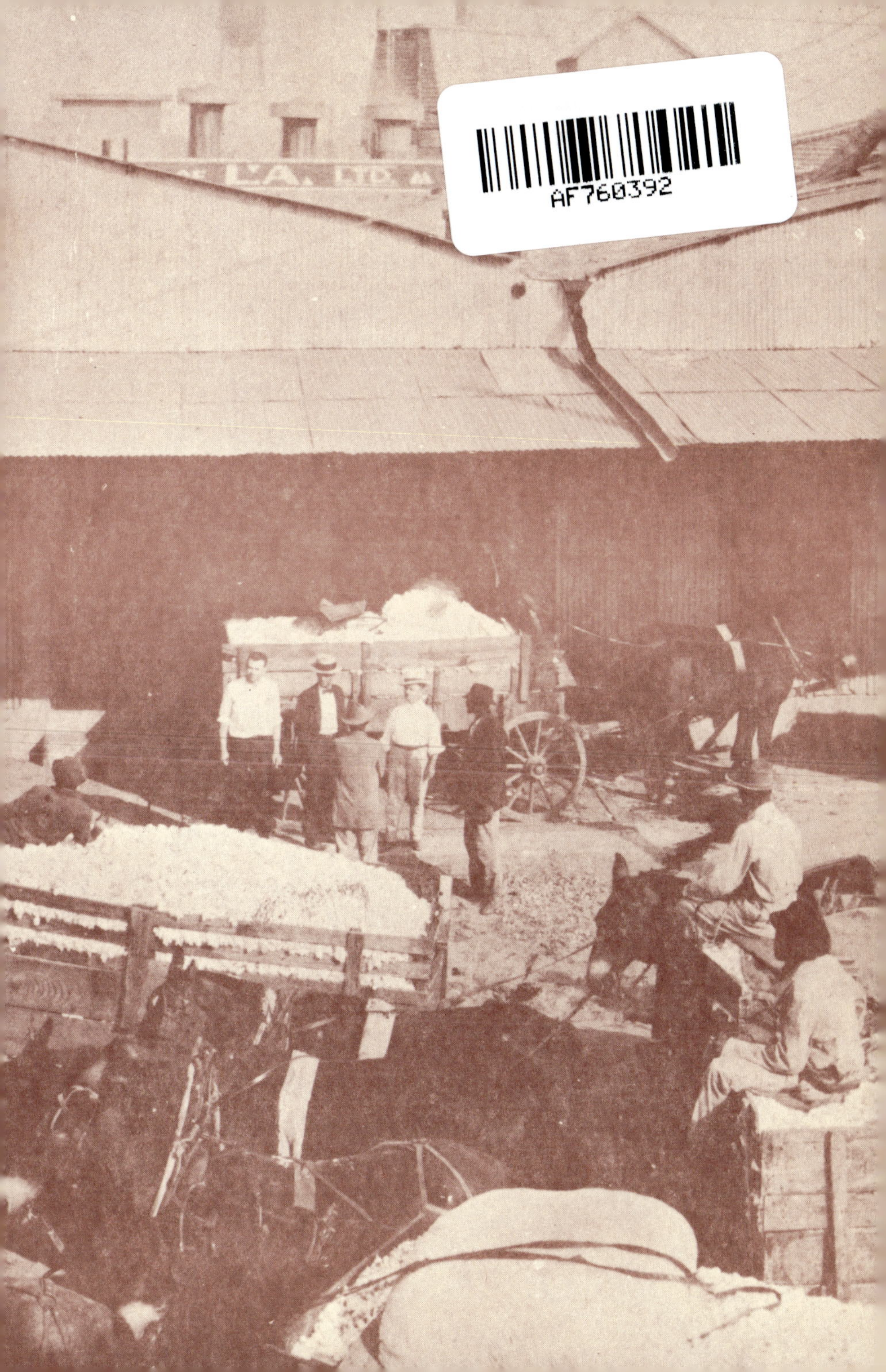

DO YOU WANT ME TO DO ALL THAT AND PLOW TOO?

Dan P. Logan, Sr.

To Gladys Tooke Barnette
with many Happy memories when you lived here
Hope you enjoy reading this [illegible]

Buddy Logan

30

DO YOU WANT ME TO DO ALL THAT AND PLOW TOO?

Dan P. Logan, Sr.

EXP

Exposition Press Hicksville, New York

FIRST EDITION

ISBN 0-682-49146-2

Printed in the United States of America

CONTENTS

FOREWORD

Dan P. Logan, Sr., who thinks of himself as a cotton farmer, wants to share with other farmers the information he has gathered from fifty years of farming. Twenty years of that time was spent in a study of how to grow two bolls of cotton where only one grew before. This is the beginning of the era of agricultural consultants concerned primarily with the feeding of the plant.

By presenting this method of improved farming and having other farmers accept it Mr. Logan will have participated in the increased yield of crops in this and other lands and will have made a great contribution to mankind. Cotton will resume its role as King Cotton.

Mr. Logan at seventy has seen farming practices that were like those used after the Civil War, as well as the most scientific techniques used in farming today. He is a friend of many of the great men of agriculture.

His knowledge of the old ways of farming should be preserved. It is hoped that his knowledge of improved methods of farming will open the door to increased production and profit for cotton producers.

N. D. Morgan, Ph.D.
Agronomist

AUTHOR'S NOTE

I wish to dedicate this book to the young men and women who have just decided to begin to farm and to those in the future who decide to make agriculture their life work. The future in agriculture is so great for those who will think for themselves.

There is always a challenge in farming, for there is never a dull day. Nature has her secrets but she will reveal them to all who will look and see for themselves. If you don't discover some of these secrets, the fault is yours. The more you discover, the more there is to be discovered. All of Nature's growth, habits and reproduction methods are open and exposed for you to see if you will. Nature is no respecter of persons and she will grade your term paper as to whether you pass or fail.

It is my hope that this book will scratch you enough to make you bleed so you can see if there is just a little "farming blood" in you. Then maybe you can work and develop a way to increase production of food and fiber, so that the many people who know nothing about farming or are unable to farm, yet are totally dependent on farmers for their welfare, may benefit. The great majority of people are totally helpless today and must be fed and clothed.

Agriculture will require the best brains in our nation because the whole future of the nation depends on people who are dedicated to food production.

All wealth comes from the ground—oil, coal, wheat, corn, rice, cotton, beans, beets, potatoes, and by extension, all food. Dollars are not wealth. Bonds are not wealth. Multistory buildings are not wealth. Data processing is not wealth. All these things represent wealth. How will the average person react to a genuine gasoline shortage? How will the average person react to a genuine food shortage?

You, as farmers, have a great responsibility—for you produce *true wealth*.

ACKNOWLEDGMENTS

I wish to express my appreciation to those who encouraged me to write this book of my ideas and findings, based on my fifty years of experience in growing cotton:

To Colonel Val Irion, attorney, friend and advisor, who showed faith in my ability to complete the book and offered me spiritual and material assistance.

To Dr. Nevin D. Morgan, retired professor and one of the most knowledgeable and best-known authorities in the South on plant food. He has spent his life teaching and encouraging farmers to feed their crops a complete food. It was he who kept telling me over the many years of our association that there was too much nitrogen used at the wrong time for the cotton plant to produce its maximum.

To Mrs. W. A. Codding who offered a lifetime of experience in literature and publication, and furnished much valuable advice and wisdom.

To Mrs. F. C. Boston, author and friend, without whose

direction and wide knowledge and patience this book could never have been completed. She typed the manuscript over and over again, arranging and rearranging the contents so that the subject could be easily read and understood.

To James Dillard, who took over the responsibility of looking after the daily details of running my farm and let me study the plant and its actions and write it down. I could not do all that and plow too!

PREFACE

The purpose of this book is to discuss what I believe to be the truth. I am a cotton farmer. What I believe to be true today about agriculture may or may not coincide with your belief. It is your privilege to disagree. Of course, no individual has the whole truth, for if he did he could then cause the cotton plant to produce such a bountiful crop that we would not believe it possible.

Farm machinery has become specialized and efficient. The plant breeders have given us good seeds, the chemists have developed useful food supplements, insecticides and herbicides. The farmer must still make the decisions as to the time of planting and when to give food and water to the plant.

We farmers must learn to apply in a practical way the ongoing agricultural research, so that the human beings of the world will have enough to eat and some to spare.

For fifty years I have farmed this same land I am now farming. I have searched for ways to increase production. I have learned many things that won't work and a few that will. Sometimes I made the same mistakes year after year. I didn't know what I had done wrong. When I found that I could not remember each detail, I began to write a daily record. For a period of fourteen years I have written about the day-to-day happenings as I saw and understood them.

It is hoped that these observations will have deep meaning and significance for all farmers who seek to observe more closely what has happened, and will add to the truths that are necessary for farmers to use daily. I believe it is up to the farmers to follow the truths of agriculture so that we can add to the production of those things city folks, as well as country folks, will shortly need.

Let the truth shine forth because it might help those who wish to do a better job with their farming from now on!

I've learned that nature's ways are right and farmers prosper by following them. Those who don't follow nature's laws stumble and fall, because they ignore the fundamental truths that never vary. May those who are wise understand what is written here and may they take it to heart, and prosper.

DO YOU WANT ME TO DO ALL THAT AND PLOW TOO?

COTTON—THEN AND NOW

Your age as a cotton farmer can be determined by how many of the following terms you can identify. You grade yourself. These were things we dealt with every day: hame string, middle buster, double shovel, Georgia stock, Gee Whizz, buck head, clevis, single tree, double tree, back band, head stall, snapper bits, throat latch, eye bolt, nose twist, barr off, monkey wrench, tap wrench, color pad, blind bridle, twisted buck, loggy head, knee pad, sweet back, trace chain, jack, listing, gee and haw, tire shrinker, lye hopper, fellows, hounds, rocking bolster, coupling pole, froe, cradle, shake, mule skinner, bull tongue, king pin, breast yoke, plow lines, spring seat, side harrow, Frezno, slip, wheeler, state chain, breast chain, maul. . . .

Today it is not necessary for young farmers to know these terms and lots of you older farmers never had to use them either, but they were farmer's terms forty years ago and on back.

My first crop was in 1928 and this crop of 1977 completed my fifty years on the same land. In 1928, changes in farming

The cotton harvest isn't what it used to be! This old photo, found in the Federal Compress files, shows the harvest in Grenada, Mississippi, circa 1900. The whiskered gentleman to the right of the wagon is the buyer; the man under the scale is the county-elected weigher.

methods had just begun in a small way, but generally people who farmed used exactly the same methods that had been used fifty years before. Some of the farmers I knew and worked with began farming in 1877 and were still using the same methods, so I was instructed in these 1877 methods that were tried and proven then.

When I made a mistake, I wanted to correct the procedure. I found that events happen somewhat the same each year, or every several years, so they fall into pattern. The reason we suffer from poor judgment is that we don't remember or we don't read our notes of what happened in the past. We farmers are kind of like the old goose—we wake up in a new world every morning not remembering what happened yesterday or yesteryear. In the growing season, each day's work occupies us fully, so there is little time for reflection.

I was trying to gain practical, everyday knowledge when I was studying and writing about the cotton plant. If we had temperature at certain degrees when the cotton was trying to emerge I would write the temperature down, then in a week or ten days I would note how the plant responded and write that down. When hail came and stripped the cotton, leaving nothing but the bare stems, I would record how the plants looked one week and one month later and then I would record the final production in comparison to an uninjured crop. When a rain came in mid-July and thoroughly wet the ground, I would record how long it took for the plant to begin throwing off little bolls for want of moisture; or after cotton had already started to suffer for want of moisture and a good rain would come, how long it would take the plant to take on new growth and begin to fruit again. If I had not written down the details, I would not have remembered

Dan P. Logan in 1936, plowing new ground with the double shovel and mules.

just what day a big rain fell and just what the crop looked like after the rain was over.

My record-keeping was a search for truth. As we wrestle with the age-old job of producing food and fiber, we realize we must not be static in our beliefs but must be open to change when additional light is thrown on any aspect of our work.

There has been more improvement in agriculture in the last seventy-five to one hundred years than there had been from 1877 back to the beginning of man's recorded time.

When I started to farm, tractors were in use but were very rare. Only a select few had the one-row cultivator pulled by two mules, and the cultivator was used sparingly.

Everyone depended on cotton for a money crop. Cotton paid for the land, bought new buggies (and later cars), bought more mules, better saddle and buggy horses and new harness for the mules. Cotton built new homes, paid for the education of the children and paid for a trip, if there was enough money left over.

There was a house where the tenant lived on about every thirty or forty acres. It was located usually in the middle of the block of land that the tenant farmer worked. Almost all the houses had a long front porch with usually a dog trot dividing the rooms on either side of the house. Sometimes these dog trots were boarded up to make the house warmer in the winter. Heat came from two or more open fireplaces, where wood was kept piled on during cold weather. Everyone wore long cotton underwear during winter, so there was no use getting a house stuffy-hot as we do today. If you wanted to get a little warmer, you just stood in front of the fire. When one side warmed up, you turned around.

Most of the houses had a horse lot with a shed for the cow and mules to stand under. There was a corn crib with the front opening out of the lot. There were chickens with some frizzleys scattered among them for good luck, a few turkeys, several white geese, and usually a flock of guineas, plus several hogs. The dried peas were put in sacks, tied with wire, and hung to the rafters in the barn to discourage the rats from eating them. The cushaws and pumpkins were covered with ears of corn in the corn crib to keep them from freezing. The sweet potato bank was usually close by the house. There were usually one to three dogs of Wonder breed around the house. These were used to tree 'possums and catch rabbits.

The idea of going hungry was unheard of. There was always a crib of corn and a big bank of potatoes, with gallons of ribbon cane molasses made from their own sugar cane patch. Those of you who have never eaten a sweet potato roasted in the ashes of an open fireplace have missed one of the finest flavors in the world. Louis XIV would have given up part of his kingdom for one more taste. The only thing that came near surpassing this delicacy was corn bread baked in a big skillet, with a lid on it covered with coals of fire. When the bread was thoroughly done, a piece was broken off and stuck in a plate of ribbon cane syrup and washed down with buttermilk just churned. People who ate this never had rickets, neither did they suffer from malnutrition. The only underfed around the house were the hound dogs and that was because the children ate up all the food.

Most of the sharecroppers had been on the farm for years and most were very honest and upright. They were fine Christians and good farmers. During the depression, when no one had any money, the owner of the plantation I worked for put a mortgage

A typical plantation scene from the 1930s, showing the tenant farmer's house and cotton fields ready to pick.

on his farm with the Federal Land Bank in order to continue to farm and feed all the people on the place. No one ever went hungry during those years but we all worked and raised food and saved it to live on. If someone had a disaster such as sickness or his house burning, the others would pitch in and help him get back on his feet.

All the people attended the same Baptist church and they had a good board of deacons that looked after the physical and spiritual welfare of the members of their church.

Each plantation had a baseball team and they played each Sunday evening, between the levee and the Red River. The deacons were allowed to sit on the levee and watch the ball game because their job was to keep their members from defiling themselves by playing ball on Sunday.

Matt Lee was a fine player and he loved the game very much. He would be turned out of the church every year when ball season came around because he just could not help playing. He would have to join the church again when the ball season was over. About 1930, they got Matt Lee up before the board of deacons to turn him out again. He begged and pleaded with them not to do it again, because the season was over now for another year. He explained that they had turned him out every year for the past four years and he didn't want them to again. The deacons were determined to carry out their work despite Matt Lee's pleas and tears. Matt Lee finally got mad and stood up and said defiantly, "Now Brother Cy, Brother Howard, Brother Phil, Brother Jim, Brother Po Jake, Brother Robert, I have asked you not to turn me out of the church and you are just going to anyway. Now I know something on everyone of you and if you turn me out of the church I am going to tell it!" Amid many tears and

Didda's dear friend "Big-un"

begging their pardons, Matt Lee kept his membership that year.

When I came to work here, Mr. John B. Adger, my uncle, told me that he would tell me what I was to tell each cropper to do, but there was one cropper, Frank Marshall, on the plantation that I was not to bother. I was to let him alone and let him do whatever he wanted to do. He had found that Frank was the best farmer he had ever seen in his life. Frank always made much more corn and cotton than anyone else on the farm.

I, too, found out that Frank's crops were always superior to all others, not only on this farm but all other farms as well. I was determined to find out how it was done. I would ask Frank why he plowed the way he did. He would stop his mule, lean back on the plow handle and say, "Mr. Dan, I came to this country from South Caliner with Mr. Bill Belcher and we settled out in Bossier Parish at Midway. Mr. Bill Belcher was "gived up" to be the best farmer in Bossier Parish and that is the way Mr. Bill Belcher would have done it." I did not know Mr. Bill Belcher, for he died years before I was born. I did know the Belcher family. I drove over to see Mr. Norwood Belcher, who was an old man, and told him of Frank Marshall and his ability to farm. I said that Mr. Bill Belcher must have been one of the outstanding farmers in that whole country.

Mr. Norwood Belcher listened to my story and then laughed and laughed. He said, "Papa and Uncle Bill were brothers and they farmed together and as far as Uncle Bill being a good farmer, I have never heard it before. They farmed lots of land, but they never did one thing in the world except hunt and fish the whole time. They had hands farming their land and their crops were good but it was not Uncle Bill's ability that did it."

I came to the conclusion that Frank Marshall had a God-given ability to farm and he was laying his decisions on Mr. Bill Belcher.

As I watched the cotton plant grow and develop from year to year, I kept wondering what made it produce, or what made it throw off fruit and would try to decide what action to take. I found myself wondering what Mr. Bill Belcher would have done and kept wishing that I could talk to Frank Marshall and ask him what he would do.

There is so little we know about the cotton plant that has been King. It can promise us so much and give us so little, and can promise so little and give us so much. Those of us who make our living by growing it really don't know what to expect. Our friends in town can't understand our statement to this effect.

The amount of rain we get or fail to get oftentimes changes the whole outcome of our crops. It affects our production so much we farmers oftentimes don't do as good a job as we should. We just say, "Well, it will rain and bring the crop out." If a farmer plants a certain way or cultivates a certain way and if it rains at the right time he will be a wise man. If it does not rain at the right time, he has done a stupid thing and should have known better. It boils down to this simple fact—whether a farmer is wise or stupid depends on that rain and he has no control over that. The only difference between a smart man and a "durn fool" is one rain.

During the early nineteen thirties when the government began efforts to get agriculture back on its feet, Mr. C. C. McCrory, Caddo Parish County Agent, was explaining the program the government had worked out to a fine old colored farmer in Gilliam named Jim Johnson. Mr. McCrory explained in detail to Jim and Jim would say, "Yes sir, yes sir," often. When he completed his explanation, Mr. McCrory asked Jim if he understood it and Jim replied, "Mr. Mac, I want to ask you a question. Do you want me to do all that and plow too?"

When I decided to write all this down, I wondered how many farmers could do all this and plow too. Sometimes we wonder if it is not just as well to plow any kind of way and plant any old way and let lots of weeds and grass grow all over the place, plant any old seed for if it is a cotton year every one makes a crop, and if it is not nobody makes a very good one. It doesn't seem to make too much difference whether one tries or not. I'd like to tell a true story to bring out this point.

It used to be my custom to order the marriage license for the boys on the plantation when they got married. I kept the marriage license blanks in my desk. Paul Matt Howard came in and said he wanted to get married and told me what a fine girl he had got. She lived on Possum Walk Plantation which joined this place. In fact, he told me, she was the finest girl on Possum Walk. The license was ordered and delivered to Paul Matt. In about four weeks he came back to the Commissary and brought the marriage license back and said he wanted to get it changed. He had decided not to marry that girl, but was going to marry another girl from Possum Walk. He said he wanted me to send the marriage license back and change the name of the girl he was going to marry. I told him that the clerk of Court would have to cancel that license and issue him a new license and charge him $3.00 for the new license and probably charge him $1.00 for cancelling the old one. He said, "But Mr. Dan, I ain't never used this license. All I want that man to do is scratch out that

Cotton bolls begin to open up about September 15.

Two weeks later—cotton picking time!

girl's name and put the other girl's name in its place." I explained that a marriage license was a matter of record and had to be in a big book. He said to tell that man that if he was going to charge another $3.00 for a new license just to send that other license on back and he would marry that first girl. He said, "To tell the truth, Mr. Dan, there ain't $3.00 worth of difference in them gals."

You readers who have raised cotton know that lots of times there doesn't seem to be any difference in production regardless of what we do. There's not $3.00 worth of difference in our methods. There are many things that happen that we have no control over seem to be adverse for the plant to make a good crop and the plant makes a fairly good crop anyway. It is kind of a mystery plant. The only ones who know all about raising cotton are the ones who have never raised it. The experienced men who have been successful know there is no telling what a crop will do. The people who do the best job and have a high lint average will tell you they do not know but are trying to learn. Most of those who don't know will tell you a maximum yield has been reached and you need not try any more.

In my fifty years of farming I have observed and tried various methods to increase production. A few will work sometimes—many won't—but I have had lots of fun observing them all.

Notes for June 13, 1976, say: "We are trying to work out a practical way of feeding the plant so that those who come after us may benefit. It's not an impossible dream to increase our production. It seems to be just plain hard work. If I am not too badly mistaken, this method will work every time. I will ask you to find the faults with my methods for I believe it will stand the test of time."

DRAINAGE

In dividing these ideas into sections maybe it is best to start off with drainage as this is the most important part of all to farming.

There are two types of drainage, surface drainage and subdrainage. Surface drainage keeps water from standing on the surface of the ground and subdrainage keeps water low enough under the ground that the roots won't have to stand in water. The cotton plant will die if its roots stand in water.

Roots of most plants (except rice) cannot grow in water. The roots want moisture but they also want air. If air is excluded from the roots by having the ground thoroughly saturated with water they just will not grow.

A low place in the field where water stands or does not drain off well never produces a good crop and it's the most expensive part of the field to work. It will not produce a good crop one year in twenty.

To make the maximum crop each year a field should have

My granddaughters, Beth (left) *and Susan Logan, ages thirteen and sixteen. The girls do an excellent job of setting grade stakes on land to be formed. The job is not difficult and it pays off in increased production; it gives them something to do and take pride in.*

a slope on it of three to one or three-tenths of a foot fall to every one hundred feet of row. Rain or irrigation water will not stay on a slope like this long enough to water-soak the land. This slope also will not water-soak or waterlog during long spells of rainy weather.

One of the problems all farmers have is the land getting water saturated, causing the crops to lose the nitrogen in the ground and stop making. By having the proper slope, this condition will occur so seldom that it won't be called a problem. Level land won't produce a crop one year in twenty. Land with a one-tenth slope will only produce a crop sometimes. Land with a two-tenths slope will produce a profitable crop about two out of three years but land with a three-tenths slope will produce one almost every year. I like four-tenths of slope on stiff land because it produces even better.

A banker who lends money to a good farmer whose land has a three-tenths slope has got a sure loan.

All crops need moisture and well-drained land will hold more moisture for a longer period of time than land not well-drained.

LAND PREPARATION

The reason for breaking land is to kill the vegetation so the new crops can be kept clean easier and also to mix the residue of the previous crops into the soil, so that decay will be faster and furnish food for the following crop. It seems to me that in order to kill the Red vines that grow from the roots, these old roots will have to be plowed out. I doubt if mixing up this residue of the previous crops aids the following crop. It seems to me from observation and from what I get from research that crops will make just as much, if not more, if the ground is not broken with a plow. Just leave the residue of the previous crop on the surface and keep the vegetation under control. A row should be built up as high as possible in the same place the row was last year. This row should be dragged off and the seed planted on top of this high row which will warm up faster and dry out sooner—two important factors that cause plants to grow fast and early. A crop usually makes more if it gets off to a good early start in the spring. These young plants need phosphorus to help them develop early and phosphorus does not seem to be released for the plants in cool wet soil. This high bed helps it be warmer and dryer.

Most of the preparation should be done to get the land in the right shape to drain and then set the rows up. The extensive land preparation every year is not needed. After the land is formed right, you really do the land harm to do extensive plowing and disking each year. The least you can do to the land after it is formed right is the best, I believe. For an example: After cotton has been picked, cut the stalks and run a disk over the field lightly, then hip up on top of the old stalks. Then come in with a Big Ox plow with a subsoiler in each middle and break up the hard pan that was caused by the traffic over it all the year. This subsoiler should go as deep as necessary to break the hard pan. Deeper than that does no good. Nature seems to do a better job along this line than we do with all our large horsepowered tractors.

It is my belief that the disk is one of the great enemies of the farmer for we use it too much. It packs the ground over which it runs but makes it pretty to our eyes. So much of this extensive breaking is not necessary for the best interest of the crop. The crop needs to be kept free of competition if possible. How?

PLANTING

It's usually best to plant any crop early enough for it to take advantage of all the season of that year. The time to plant each crop depends on the temperature that will sprout the seed of that crop. Corn and milo will sprout in colder soil than cotton or beans. We will discuss one crop and others can be substituted.

For cotton the bed has to be around sixty-five degrees before the plant will do anything at all. Some kinds of land warm up faster than others. Black color absorbs heat from the sun and white repels the sun's heat. The blacker the land the more heat it will absorb and the earlier it can be planted. The white, light colored land absorbs heat slowly so it must be planted when the weather really warms up.

Your black land or darker colored land can be planted and will grow off much sooner and faster than the light-colored soil. All crops should be planted on as high a bed as possible, for the high bed will absorb more of the sun's rays and be warmer and drier than the flat land.

It is necessary and important to get the crops made as quickly as possible and then get them out before bad weather sets in.

Keep this in mind always when you farm, it doesn't make any difference how much you make if for weather reasons or other reasons you don't get to gather it! The weather in this midsection of our nation does not always cooperate with us. We have to make the crop as early as we can possibly do so and then get it gathered as soon as it is ready. Sometimes picking cotton just one week earlier will make lots of difference. This high bed often makes one or two weeks' difference.

Most all the cotton seed varieties are good but it will pay to put seed treatment on them to assure that diseases don't stunt the young plants.

It is my belief that systemic treatment is very beneficial to the early setting of bolls. It should be used for best results. Early bolls will pay off well.

When the seeds are planted on this high bed, they will come up in a few days after planting. The plants should have from one to three stalks per foot. How to get this plant population?

CULTIVATION

This word cultivation is one of the jobs that we could do without if we could find a way to keep the vegetation down without it. The only reason you cultivate is to kill weeds or grass or other vegetation.

We pack the land between the rows with our tractors. Lots of times we turn up new soil that has new weed and grass seed that would not have germinated if the soil had not been turned up.

Every cotton root we cut with the plow injures the plant just a little. If we could find a practical way to keep from cultivating we would be better off. How?

The purpose of cultivation is to remove unwanted plants that will compete with your crop for food. On July 2, 1975, my farm notes advise: "Plant on a high bed after the ground has warmed up. Have all vegetation killed by mechanical means or by chemicals. Cotton planted on a high clean bed grows off faster and is not affected by big rains as much as on a low bed or planted flat."

Three days later I wrote, "If you do get in the grass early it is better to kill cotton, grass, and all, with acid or Paraquat and plant again if you can do so by June 1. You will make more cotton and make it cheaper than if you try to work this old cotton out. Use as little herbicides as possible to keep it clean because herbicides stunt and hurt the plant each time. No herbicides will be necessary if the plant comes up clean and then oil is used twice—five gallons per acre each time. This oil will not stunt or injure the plant if it is kept out of the bud of the plant. This herbicidal oil has to be used when the plant has four leaves and then again two weeks later, whether you think it needs it or not. Sometimes the weather won't allow you to use the oil on time and then you will have to come underneath with other material.

"If you keep the vegetation down, the plant will grow better by not plowing it. It is not necessary to cultivate for the plant to do its best. I really think we hurt our production every time we plow. Rain usually does not injure the cotton plant during July and August. If it is clean when July comes, the plant will stay ahead of the grass because it really grows during these two hot months. However, it is hard to keep the bugs off during these months when it rains."

I learned of Dr. Edward C. Greco's work on hyacinth control and asked him to talk to some of the farmers here in Gilliam about the control of noxious weeds. The tie vine was in my fields making it difficult to machine pick my corn. Dr. Greco helped in many ways. We are doing now what he suggested then.

About 1930 or 1931 I started taking soil samples. I continued over the years trying to learn more about the condition of the soil. In notes for May 25, 1973 I wrote, "While putting out moisture blocks yesterday I found the ground very hard down to seven inches and it was difficult for the probe to get through this section. The cotton plant was small and not growing. Below eight inches the ground was soft again. This was the situation where the fields had been disked. Where the fields had not been disked the probe would go right on down and the cotton plant was growing and vigorous and looking fine. The disk is an enemy to the farmer most of the time. When the land is perfectly dry it is all right to use it but it will pack the ground where there is moisture."

Another warning about how the ground can pack was in notes made on July 4, 1974. "After the long siege of rain we plowed where we could. Where we plowed and the ground was crusted over on top and wet underneath, after the ground dried it was very hard and the plow the second time would not go into it. To remedy this, plow the ground the second time as soon as it dries but before it gets hard. Where the ground has never been plowed, it will not get hard. I think this will work very well."

The following year on July 27 I wrote, "We made some mistakes with the crop and I wish to write about it so we won't make these mistakes again. We could not plow the land and clean it but we did not kill all vegetation with chemicals before planting. We had to kill the vegetation in the cotton after the cotton came up. This was bad because it not only costs more to do it that way, but we killed lots of cotton and stunted the balance. We still have tea weeds and careless weeds and high Johnson grass spotted all over the field. I cannot help but try and get these out myself and this takes time. You cannot get hoe hands any more. It hurts the cotton to stunt it with chemicals besides making it later to produce."

Follow the instructions given on the many chemicals used to control vegetation and keep up with the environmental effects to know which are best for you to use. In 1976 my notes say, "Where we had used Treflen correctly the crop was fairly clean but don't depend on Treflen alone. Break the land well. Put it upon a high row and keep the top of this row clean some way."

WATERING OR IRRIGATING

The reason for irrigating is to keep from taking a terrible loss when short periods of drought occur. Drainage is the most important thing we can have for as a rule we have too much water. When the land is fixed correctly we have perfect drainage. In case of a drought we can supply water efficiently and inexpensively. Preparing land and getting ready to irrigate is the cheapest insurance you can take out.

The land I farm is like the land around Lubbock, Texas, or the Palo Dura Canyon soil in Texas. It came down Red River over eons of time. On the river bank, the land is fine sandy silt. The stiff land, like that around Lubbock, is dark-colored and cracks open and needs lots of water in July and August so it won't burn up.

This area was occupied by the Caddo tribe of Indians. They had lived here for centuries. These many lakes were drained after the Red River levee system was started in 1893 and completed in 1908. Water brought all this soil here and the soil is approximately sixty feet deep. After the surface water is drained off and the water table in the soil is lowered, this silt will produce beautiful crops of all kinds. There is one draw back. This soil does not have the capacity to hold enough moisture to carry it over two hot dry months of July and August. Sometimes—almost every year—the crops that don't get additional water during one critical time during these hot months will suffer terribly and not produce much, just for the want of sufficient moisture for a few days.

Unfortunately, we can't predict when the moisture will fall from the sky, so we have to add this moisture before it is needed, because you can't water all the acreage in one day. We have to begin watering before it's needed so that none of the crop suffers from lack of moisture.

If the moisture is put on the crop properly, it will not injure the plant in the least during the months of July and August. The plant seems to be able to use all it can get during these months. There is a possibility that it can rain so much during these months that cotton will be injured, but this will seldom happen. I know you can soak the land with irrigation water during these months and your crop will be cut way short. I did this. I thought at the time the crop would be injured and took slides. When the cotton opened I took more slides and the waterlogged cotton yielded less than one-half.

This soil can be watered so it will produce a fine crop every year but it has to be watered the right way. You must put the water on fast and then take it off. The moisture meter will show you when to water again. There will be very few years when you will have to irrigate more than once or twice. Before learning how to use the moisture meter, we were just guessing. The use

Inserting Moisture Meter box to determine water level in soil

Meter reading of moisture level taken twice a week during growing season.

of the moisture meter to tell when to irrigate has taken the guesswork out of watering.

To irrigate for maximum production, it is important that water be applied before the crop is injured by stress. When the plant wilts severely, it has already been injured. The moisture chart will show you before this happens.

In the spring Tolly Burns, who had just retired from teaching agriculture at North Caddo High School, offered to conduct a study on where the moisture is in the soil all during the cotton-growing season of 1977. This was something we all should know and I jumped at the chance of his making such a study.

This is the description of how he went about the study. The method is simple yet gives us the information that we had always wanted. This study answers the question of when to put water on the cotton plant and when not to water.

Tolly Burns used the soil moisture meter that I had used for many years. It was the soil moisture tester made by Delmhorst Instrument Co., Boonton, New Jersey, Model K.S. We used the cylindrical soil blocks made by them with three foot leads. They were set at eight inches and eighteen inches depth in the row of cotton after the cotton came up to a stand. A regular soil probe was used to make the hole. A one-gallon plastic jug of water was carried along and the cylindrical blocks were lowered in this to get the block thoroughly wet before it was inserted into the hole and pushed down to the bottom with a stick. Fine dirt was put into this hole on top of the block and enough water was poured in to cause it to completely fill the hole with dirt. A flag was placed so this spot could be found.

All that is necessary then is to insert the wires from these blocks and read the meter at regular intervals. This method will tell you exactly where the moisture is in the ground and how much is there at each reading.

This year is the first year that I kept these records or read these blocks at regular intervals. They definitely will open anyone's eyes who cares to look at what the moisture condition is in the cotton or bean field. I can tell when cotton or beans need water for those blocks register from saturation (that is 200) to where there is no moisture available (0). I do not mean that this is the only method by which moisture can be measured, but I mean that this way is simple and practical and cheap and anyone can do it. My sixteen-year-old granddaughter takes these readings much of the time. It is worth anyone's time and expense to know what the moisture condition is in their fields, whether they can or will do anything about it or not. "It seems as though, from this year's results, if there is no rain, bayou land should be wet every seven days and stiff land each twelve days for best production. If you get a rain, you just save that much water and the expense of pumping it. The plant will need this much water in July and August. I have the complete results and movements on my nineteen cuts summarized."

You can set as many blocks as you think best over your farm and take the moisture readings twice each week. With this information you can tell when to irrigate and the blocks will show how deep the rain you did get went into the soil. Some types of soil will take up water much faster than other types of soil. This method opened my eyes to how fast moisture dries out of the ground in the summer months of July and August.

I have thanked Tolly Burns for bringing this moisture truth out so any farmer can tell when irrigation or water is necessary. How we use this information is up to us, but we will know where the moisture is.

There is a copy of some of my moisture readings in case you wish to take a look yourself. These findings have been drawn on a graph to show the movement of the moisture in the soil. There is one rough copy that I made before the graphs were drawn that shows plainly where the moisture was. Some farmers may like to look at a graph and some may like to look at the figures. They both tell the same simple story. (See pages 52-54.)

It is important for a farmer to know when to apply additional water to his crop and when not to. There are practical ways to irrigate Red River land in Louisiana. Moisture must be furnished with as much knowledge as we can get. We should know how much moisture is available in the top eight inches of the soil and at the eighteen-inch depth. Do not soak the fields. Put the water on fast and take it off fast. Do this in three or four hours if possible. Just pass the water over the top and then move on for there might be a big rain tomorrow night. Come back in ten days, or whenever the moisture blocks indicate if there is no rain.

Irrigating is a timely job.

The cheapest and easiest way to water crops is to run water in the furrow with two-inch siphon tubes taken out of a temporary ditch at the head of the field. The gated pipe and sprinkler systems are slow and expensive.

A pad must be at the head of the land and should be at least six inches higher than the field. This pad should have from one-half to one-tenth fall per three hundred feet, falling away from your water supply. This pad is a turn row except when it is opened up and water is run down it. If a ridge is available water can flow in both directions.

The field should have slope enough for water to run down each row and be distributed by gravity. A swag in the field causes trouble and will cost you money, worry, and water. The rows should be as high as possible. If flat planting is practiced, build the row up with something.

The force of the water from the well or bayou must be killed before entering the ditch or it will wash everything away and it can't be handled. Do not use a concrete distributing box. I did at first but found it's too expensive and requires too much labor. I use a metal box with openings leading to different drainage ditches.

To do an efficient job of irrigating and to expend less labor, use two men and a tractor with a tube-carrying device on the back carried by the three point hitch. This device should carry at least one hundred tubes, all the long and short sticks, and all the canvases needed. The two men are to set every canvas and to set everyone exactly alike. They are to practice until they can do this correctly. They are to set these canvases where the management sets a red flag. When the tube starter takes a tube out, he should throw it over the ditch for the tractor man to pick up. There will be a tube at the right place put there by the tractor man. The long two by fours should all be the same length and painted the same color. The short sticks should be all the same length and painted another color. There should be a surplus of canvases and all of them in charge of the tractor man.

Most of our troubles come from improper setting of the canvases and improperly made ditches. Most of the work in starting tubes is having to carry the tubes forward. I made a tube carrier as mentioned and I use black plastic to stop the water and black plastic pipe to carry the water from the ditch into the rows.

Fourteen- to sixteen-year-old school boys are best to look after the irrigation. After they are trained, two apt boys can look after two hundred fifty to three hundred acres of land if the systems are installed correctly. The boys can run the motors if the power is electricity. If gasoline motors are used for power, the motors will have to be handled by someone else. The systems should run twenty-four hours per day, day after day, if laid out properly and common sense is used.

A well making one thousand gallons per minute will irrigate one hundred acres of one crop if it's handled wisely. Don't wait until the crop is burned up before starting to water. The saving of one crop of cotton or corn is sufficient to pay the entire cost of installing an irrigation system. From then on whatever is made is a return on good judgment.

When you have water, the rate of fertilization for maximum production is increased a great deal. Never allow the plant to stop growing for want of water. When the leaves wilt severely the plant has already been damaged.

On August 5, 1964 the temperature had been 104° and the cotton was opening fast. We had some non-irrigated cotton that year. It burned up. The cotton was too small to pick. We got a stripper here and got seventy bales of cotton from one hundred acres of land.

On July 20, 1969, we were watering every sixth row to save water and labor and get over the field faster. My notes for that day "Can all the land be done this way? A man landed on the moon tonight. Wonder if I can get a reservation to make the trip. I wonder if the moon has heaven dust on it."

Notes for September 26, 1969, read "Ray Smith and Dr. Mike Svoboda, Field Research Specialist, came by on this boll rot study and I saw boll rot when it had not rained but my irrigation water was put on wrong. By putting water on correctly,

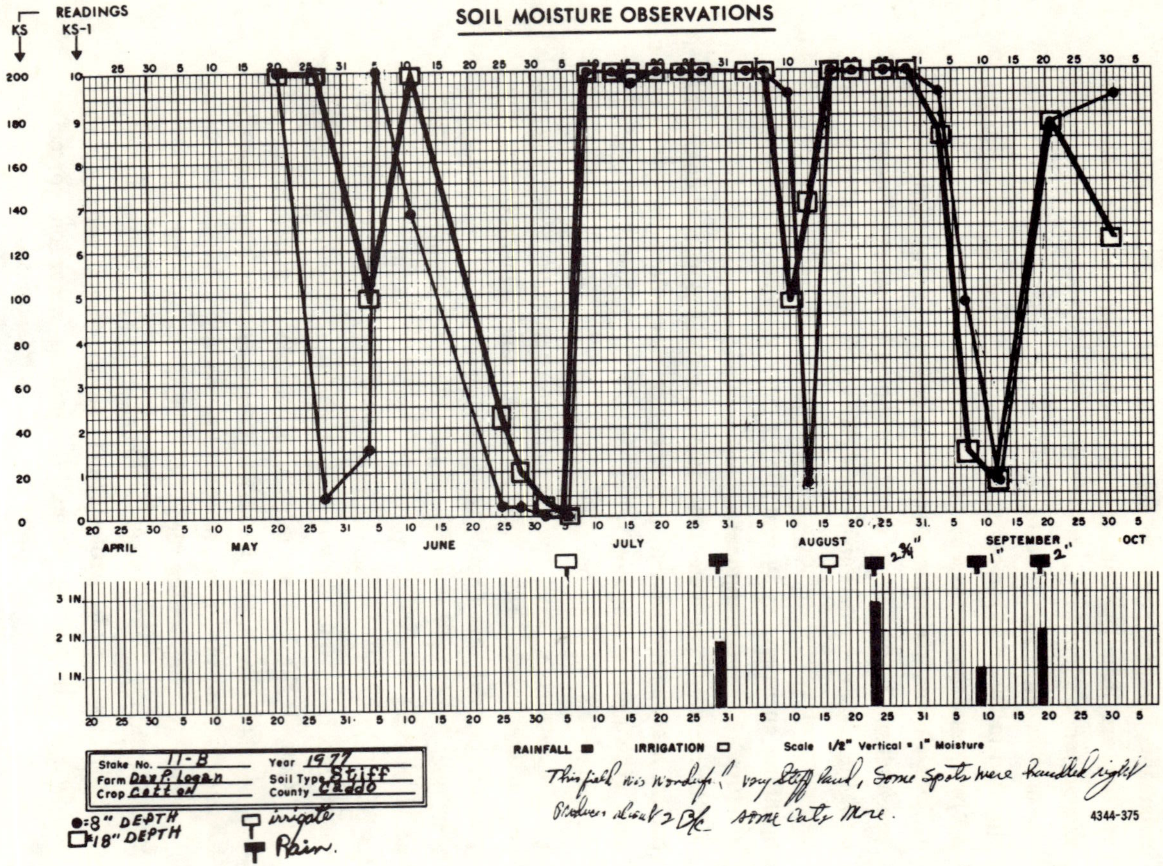

Graph chart showing running record of moisture readings during growing season.

enough bolls can be saved to pay the entire labor of irrigating."

Arrange all irrigation ditches so that you can ride past to inspect the watering.

A note of caution found in July 4, 1972, records should be repeated today. If the land is water logged from improper irrigation the nitrogen moves out of the soil and the plant starves for it. If this does happen you can fly on nitrate nitrogen or foliar feed with urea. Both ways are practical but, caution, if too much nitrogen is applied the plant will get to growing and your increased production will be defeated."

From my notes September 1, 1976, "I have just returned from over the mid-South and most of their crops are bad. No rain in most places since July 4th. Our crop here is wonderful due to pumping water and feeding right. Anyone can do this if they wish. This is no secret."

The following table shows a record of my field notes for one day's moisture readings. I take the readings every three days and the notes are for my permanent records, as shown in the graph above and in an easier-to-read form following these notes:

Saturday, August 13, 1977

Plot No.	8″	18″	Remarks
1	175	0	Dark Green 30 in. high. Burned up
2			
5A	5	45	Green 36 in. high. Making fine
5B	0	0	Green brown 30 in. high. Burned up

SATURDAY, AUGUST 13, 1977

PLOT NO.	8″	18″	REMARKS
5C	2	3	Green 48 in. high. Growing and making
5D	1	6	Green 40 in. high. Growing and making
6A	22	35	Green 46 in. high. Making fine as can make
6B	12	10	Green 40 in. high. Could hardly be better
6C	35	18	Green 50 in. high. Wonderful
6D	5	8	Green 50 in. high. Wonderful
7	190	8	Green 50 in. high. Making fine now
13	20	3	Green 45 in. high. Irrigated for third time. Making fine
9	10	4	Green 45 in. high. Irrigated third time. Making fine
10	200	200	Green 46 in. high. Just irrigated, making fine
11A	20	35	Green 45 in. high. This seems to be perfect
11B	140	15	Green 40 in. high. This seems to be perfect
31	185	32	Green 38 in. high. Making fine
33	0	0	Brown green 36 in. high. Blooming all over and out of top. Can't water
35	200	195	Light green 50 in. high. Just irrigated and seems so fine

This is my own method of keeping chronological records of each cut's moisture level during the growing season. I find it easier to read than the graphs, although the information is the same. This reading is for Cut 6A, stiff land, cotton injured early by herbicides and very small at first. Produced two bales of cotton per acre. Even though the cut got off to a late start, it made a fair crop. By taking extra care with herbicides and proper feeding, it should do better in 1978. The readings are for eight and eighteen inches, respectively:

June 6 200-200	August 6 200-200	
June 11 200-200	Aug. 10 130-55	
June 25 32-200	Aug. 13 22-35	
June 28 10-195	Aug. 17 5-25	
July 2 5-150	Aug. 19 4-25	
July 6 0-10	Aug. 20 195-195	just irrigated Aug. 23—2¾ in. rain
July 9 0-0	Aug. 25 200-200	
July 13 0-0	Aug. 29 198-199	

July 16
0-0

July 20
195-200 irrigated

July 24
200-200

July 27 rain July 31—1¾ in.
95-20

Aug. 3
200-200

Sept. 3
170-170

Sept. 7
36-20 Sept. 8—1 in. rain

Sept. 12
5-13

Sept. 21
140-160 rained 2in. Sept. 19

Oct. 1
6-10

INSECT CONTROL

Insects must be controlled to produce a good yield of cotton. When insects are not controlled, they cause small bolls and shedding of squares, resulting in excessive vegetative growth of the cotton plant.

PETIOLE ANALYSIS

Petiole is the name of the stem that goes from the stalk to the leaf. It has been found that the stem of the last mature leaf on the plant has within it the same food that the whole plant has. The plant in growing takes the food from the soil and sends it up to the leaves where the sun shining on the leaf causes this food to be manufactured into the proper substance that causes the plant to reproduce its kind. When about twenty-five or thirty of these stems are broken off and sent in to the lab as a sample from the field the chemist can get an idea of what is taking place within the plant on that particular area. Analyzing these petioles will show you what food is within the plant or what is lacking and should be added.

This is what you want to know. You want to know what food is available to the plant and whether or not it has too much of one food and not enough of another. If the proportion of the food is out of balance within the plant, the plant will not make as good a crop as when it is in balance.

We might have put out what we thought was an ample

amount of potash but the plant itself is not getting it. Why? We might have included a liberal amount of phosphorus and the plant itself is not getting it. Why? Some of the minor elements might not be present that will cause the necessary chemical action to release this food for the plant to use. Why?

In this day of large acreage it's impossible for a planter to give the close supervision to his crops that is necessary. A farmer today does not hire an expert diesel mechanic but hires a top man from one of the reliable implement companies to adjust his tractors.

The job of looking after the plant food needs of his cotton plant is most important because his bank account is at stake. This takes close supervision by people who know plant food in order to keep your operation on a paying basis. The day of guessing what is needed is fast coming to a close.

I will try to show how important it is to feed the plant a balanced diet and to feed it when necessary. Some ideas to be considered:

How to cause the plant to joint close.

How to cause the plant to square.

How to keep the plant from going vegetative and growing too large a stalk.

How to bring to maturity every bloom that comes on a plant before frost limits this possibility.

How to cause the plant to shed its leaves naturally. Only a feeble start has been made toward the solution of these problems but I think a first step has been made.

There is so much that we farmers do not know. It looks as if we are going to have to depend on the chemist to interpret for us what is taking place in the plant and make suggestions to us as to what we have to do to get our plant's food in balance so we can make a profit on our farms. November 13, 1976, notes "profits have made the U. S. the strongest nation in the world. By adequate profit, improved efficiency can be maintained. By feeding the plant right your net profits will be improved. Nature most of the time is so good to us farmers that we make a profit regardless of the many bad decisions we make."

The plant knows what food it must have to make a maximum crop and always will know. The thing for us to do is find out what the plant needs and furnish that need. I think all farmers will agree with this simple statement and finding this out is what farmers must do. The better job we do in finding out what the plant wants to eat, the higher our production will be.

September 26, 1976, from my notes, "Time will tell whether what we are doing is true or not. The cotton plant reacts the same way every time to a given condition. This never varies. I have never seen this statement in print or ever heard it before but I believe it to be true."

You can analyze the soil and tell what plant food is in the soil. The analysis might show the ground had sufficient potash, but when the plant is growing it won't get that potash into the plant for some reason. The analysis of the plant itself shows whether or not it picks up the necessary food. Thc petiole analysis is a true way to tell.

When the cotton plant is twelve inches high or it has eight or ten leaves on it, the plant at this time has within it the proportion of food which will determine whether the nodes will be long or short, whether the petioles will be long or short, and whether the plant food is in correct balance for it to produce. It's up to us to try and find out what the plant needs and feed it that food.

A farmer can send his petioles off to several reputable labs and they can give him the readings that he asks for. Now just because a man has a degree in biochemistry does not mean that he knows what food the plant needs for it to produce two or three bales of cotton per acre. Men must be trained to interpret to the farmer what the petiole analysis indicates and what should be done.

Today the method of interpreting the readings of the petioles is not an exact science. Dr. A. D. Lengyel of Lengyel's Agricultural Consulting Service and Laboratory in Phoenix, Arizona, is years ahead of anyone in this field. He is always looking for improvement, for better ways to cause the plant to utilize its food for higher production. Dr. Lengyel does not claim that any of his findings are a secret. His findings are practical and will produce a handsome profit for all those who will follow them.

At a Texas Plant Food Society meeting in College Station, Texas, I heard of Al Lengyel's advisory service in Phoenix, Arizona. Sometime later, I called Dr. Lengyel and asked him to let me come to Phoenix and see what he was doing. He had a service advising farmers what to use on their crops. Dr. Lengyel said he had pursued this idea for three years and thought he was making progress. After this visit I was convinced that he was making progress. Could I fit my area and climate into his program was the question? All I could do was try. Dr. Lengyel had never been in Louisiana. He told me that I would have to be his eyes over here but that he would be honest and if he did not know he would tell me. This was in August 1969.

I had an average of eight hundred fifty pounds, more or less, cotton per acre. I could not see why we could not get more than that. I haunted all the Experimental Stations in the South and

Dr. Al Lengyels of the Agriculture Consulting Service, Phoenix, Arizona, shown in his laboratory.

they got tired of me asking so many questions but that didn't keep me from wondering. E. K. Chandler had suggested to me that I contact Dr. Lengyel to see if he could answer some of my questions, as he had gotten this idea and started furnishing this service to farmers in Arizona back in the 1960s.

I contacted Al Lengyel in 1969 and have worked with him since that date. It is my belief that the answer is in feeding the plant at the right time with food and water. Just because we make mistakes, there is no reason his system will not work when the plant is fed properly. One of you readers will find a way to do it correctly and let the others know. You will make so much money you will be hard to live with. Who that will be I do not know. Someone will do it right. It could be you, if you will carry out your ideas. Al Lengyel had an idea and has brought us this far.

It seems to me there will have to be men trained to furnish farmers this service. Today, in the whole cotton belt, the only consultant that I know of who can render such a service is E. K. Chandler of Rosebud, Texas.

Before man found out where to learn what food the plant needed, he was not sure what to feed the plant. Since man has found out that he can examine the stem of the most recently mature leaf and that this will answer his questions, all we have to do is examine this sample of the plant and learn how to apply what the plant needs when it needs it. Examining the petioles of the plant is the same as taking blood from our bodies to see what is right or wrong with us.

This sounds simple, doesn't it? It is just as simple as it sounds. But I wonder if we are not about in the same position as Jim Johnson was when listening to the explanation of the early government program. Do we have to do all that and plow too?

To explain this procedure I would like to tell a story.

There was a boy who wanted to fatten a calf to show at the annual state fair. With the help of his 4H advisor, he bought a likely animal. They figured out how much food and minerals would be necessary to enable this calf to grow and develop into a prize animal. The finances were arranged and the calf was purchased and placed in his well-built facilities. The feed and minerals were purchased and paid for all at the same time. There was enough on hand for the calf's development.

The boy and his advisor decided it was too much trouble to feed this calf every morning and evening. What was so important was that the calf had enough and the right amount of all foods. They just placed this food around his stall where he could reach it conveniently. He could eat his fill of whatever he wanted when he wanted. The calf did not develop into a profitable animal and you can guess why.

We have been treating our crops in much the same manner as the 4H boy and his advisor did the calf. We will have to change the feeding methods with our cotton if we are to furnish the food and fiber the world will need.

A man who looks after his farm, keeps the equipment going, adjusts the plows, keeps the tractor running, attends to the pay roll, learns how to apply all the herbicides and insecticides, and waits on his wife—he just naturally hasn't much time left to find out when a plant sets bolls and why it doesn't set bolls, why it acts as it does during extremely wet weather and why the squares are thrown off when no insects are present.

I keep wondering what Mr. Bill Belcher would do, and it's a pity I couldn't call Frank Marshall back and get some final instructions from him, for he always did it right.

HARVESTING

You will make more cotton if you can pick it the day that it opens. Something goes with the cotton after it is open if it is not picked. Cotton should be picked as soon as possible even to the extent of picking it three times.

In October of 1973, I noticed cotton the picker could not pick because the locks were hard. Much of the cotton had been open for thirty days or more and it was wasted. This cotton had been planted on April 13. It came up and made bolls early, then got to growing and made another crop. The early-made bolls opened. The rains came and many bolls just half-opened after that. After the pickers went over the field, I picked up all the seed cotton and hard locks in three different spots on the field. This loss was four hundred pounds of lint per acre and at sixty cents a pound that would be a loss of $240 per acre for lint cotton alone.

You will be paid in pounds of cotton if you pick as soon as the cotton opens. It is not necessary to defoliate cotton before it is picked. The pickers, if adjusted correctly, will not injure the stalks too much going through the field.

1977 completes my forty-first year with the cotton picker. In 1936 I brought the John Rust cotton picker to the Louisiana State Fair. The cotton picker consisted of one picking unit on wheels pulled behind an Oliver tractor. After the fair was over, November 1936, we had a demonstration on Mr. Loye Beene's farm, where Bossier City now stands. He had left a big block of unpicked cotton in the center of the field. He had picked the cotton from both ends and cut the stalks and disked the land. The conditions were perfect to show off this mechanical genius that people said was impossible to make. On the morning the picker was to run, school-bus loads of children and men and women from everywhere converged on this wide open cotton and started through the field. We tried to stop them but our efforts were useless. Mr. Beene and I just stood and looked (you can imagine them running through this field of cotton). One big school boy with a heavy overcoat on ran down a row of this white cotton with a leg on each side and the coat tail dragging over the top. Mr. Beene just held up his hands in horror! That great crowd did a perfect job of thrashing the cotton to the ground before the cotton picker could get to it.

In 1937, a self-propelled machine was used for the first time on my farm. I had over one hundred hands in the field. This one machine picked more cotton than they did. It was ten years before this thirteen bales per day picking was bested.

I looked at my notes and find that on September 14, 1975, I write, "If the food is right in the plant when a boll gets mature the leaf that feeds that boll will fall off. If the weather is dry the boll will fluff out and open up. If it's too damp the boll won't

John Rust, inventor of the Rust cotton picker, the first practical mechanical picker.

Present-day self-propelled cotton picker.

open, but mold and decay will set in. It will rot." And the next day I continue, "Don't defoliate, just pick. Most years we can do this if we control the feeding correctly. If it rains all the time the open cotton will rot as it did last year in 1974."

A month later I find in my notes, "Started picking yesterday on Cut 6 A. I feel like a little boy who is waiting to open his Christmas stocking. I want to see how much cotton the stalk is making. The picking machines are doing almost a perfect job. There is not as much as there would be if we wait a week but we better get what we can now for we will still be picking when the other opens up."

On March 27, 1976 I wrote a letter to Mr. Roy Douget.

Mr. Roy Douget
Manager of Gilliam Gin Company
Gilliam, La.

Dear Roy:

On December 7, 1970 I finished picking my 1970 crop.
On February 29, 1972 I finished picking my 1971 crop.
On May 14, 1973 I finished picking my 1972 crop.
On April 23, 1975 I finished picking my 1974 crop.

We have got to pick our crop earlier and I want to change from my past ways and do better.

I wish to reserve the Sucker pipe at the gin for September 15 in 1976 (this year). You can have your gin crew ready to gin that day as I will have a line of trailers waiting to be ginned off.

Yours very truly,
DAN P. LOGAN

I picked the first cotton September 13 and we had several trailers at the gin. The showers caused us to stop picking. We ginned September 15th. The first bale of cotton sold for $1.00 per pound—the normal price was seventy-five cents per pound. The crop was open.

The best way to make lots of cotton per acre is to get it up as early as practical. Keep the early bugs off. Proper balance of food will cause the nodes to be short and then furnish enough nitrogen to hold the squares and bring them to maturity during the month of August. Plan on having the plant ready to pick on September 15th by having one half of the crop open by that date. The seed cotton will be heavier then than at any other time. The color will be brighter and it will bring more money. No one has figured out just where cotton goes that has been open for several weeks. There will be the same number of bolls on the plant later maybe, but you won't pick as many pounds of cotton as you would have if the cotton had been picked sooner.

Since we sell our cotton by the pound we should try to get as many pounds as possible wrapped up in a bale. If any of you farmers disagree with me about this statement, or if you know what becomes of the cotton, please write me a letter for lots of us want to know. You can just address the letter to Gilliam, Louisiana, 71029, without a box number for we are working on getting more population here and when we get thirty more we will have one hundred.

FERTILIZATION

Our yields are going down and all cotton people in the mid-South are in trouble. Our costs are going up and we wonder why our yields can't be brought up to meet the additional cost. Why is our production not going up? Why? Oh, why? This is the reason: A crop has to have a balanced diet for it to do its best. It is just like the human body which requires proper nutrition.

Furnishing food in the proper way at the proper time is the key to the breakthrough that will increase production in cotton.

After these many years of work, trying to see why a plant sets fruit and what causes the squares to stick, I have come to these conclusions.

It is best to keep all the young squares that come on the plant. If the plant is fed right there will be a square at every joint, starting at the sixth or seventh node if the insects do not knock them off.

Oftentimes early squares are shed because the food within the plant is not in balance and the squares will abort. When the nitrogen is higher than the phosphorus, and the plant is young, those small squares will not set. The plant will grow fast and the nodes will be from six to seven inches long. As a rule there will be no early squares set, not because of the insects, but because the food within the plant won't allow the young squares to stick. This out-of-balance food is the reason for no early squares nineteen out of twenty times.

This excess nitrogen within the plant acts in the plant as does the birth control pill in a woman.

It is my belief that a cotton plant will never set an early crop until the nitrogen level is in the correct balance with the carbohydrates in the plant. The only way you can determine this proportion is by testing the petioles. Some soils will hold nitrogen longer than other soils. Most of the time it will take several years to bring the nitrogen level in the soil down. So every time a square is missing on the plant, it is not necessarily due to the insects. Often the food is at fault and we blame it on the insects.

It is important to clean the plant of all insects twenty-five days after it comes up and to keep it clean. A good insect man is needed but he can't cause squares to stick that out-of-balance food has knocked off. If the plant is fed right and the early squares are allowed to stick you should be picking early. In a few years, more and more farmers will be getting their cotton to maturity in September and October instead of November and December.

We should be able to start picking the first week in September, if the food is in balance within the plant and the insects have been under control.

This method will work every time if you can do it right. In fact, so many bolls can be set the stalk will just break down. I

Fields of properly fed cotton, ready to be picked.

Another fine crop, ready to pick.

have not a single idea about how to cause the weather to be fair so the cotton can be gathered, but this method will put bolls on the stalk.

All of us farmers have been guessing at the food needs of the plant all our lives and it's hard for us to quit. It's just as practical for us to take a tissue test to determine the food needs of a plant as it is for a medical doctor to take a blood test on us to see what our trouble is.

The food within the plant needs to be balanced. We are just beginning to inquire into the balance and to find out what is necessary. What I will advocate now is what I believe at this time to be right, but by no means is it the whole answer. It will work fairly well until a better fertilizer formula can be worked out.

In the past I have had so many tell me they knew what the plant needs and each time I found out they did not know. Their formula would work sometimes and sometimes it would not. Each of you can go along with this feeding but still be on the lookout for improvement.

Put out a starter fertilizer before planting with a small amount of nitrogen and a liberal amount of phosphates and potash, especially phosphates. Young plants need phosphorus and when the ground is cool and wet it doesn't seem to release the phosphorus in sufficient amounts.

Just as soon as the squares start setting, furnish just enough nitrogen to keep it setting and not so much that the plant will get into the vegetative stage. At this stage it is critical that the plants grow and make in the right way. This critical time usually occurs the last of June and the first two weeks of July. There are several ways this nitrogen can be furnished and sometimes all these ways seem to work. NH_3 can be put in the middle between the rows earlier and the plant roots will reach this nitrogen about July 1. Or you can use ammonium nitrate broadcast, if it is broadcast after the plant starts making. Another way is to use some form of liquid fertilizer as a side dressing.

The nodes of the plants should be kept from two to three inches long. If they start growing six to seven inches long, sugar in the proper amount can be foliar fed to change the growing ratio within the plant so that the stalk won't grow so large.

Let me pause to issue a word of caution about setting the early squares and bolls. It takes approximately sixty days from a bloom to an open boll. If you wish to pick cotton the last of September and early October, the blooms have to come early. If your plant is in full bloom on July 15 these bolls can be picked 60 days later, on September 15. The blooms that bloom on September 1 will open after November 1, for it takes longer for a boll to mature as the days get shorter and cooler. Keep this in mind and get as many early blooms as possible.

Keep the nitrogen on the low side until after the plant gets to setting its growth pattern. The nodes should be two and a half to three inches long. Then add the nitrogen and keep it in full bloom until September 1 and a good crop will be made.

Add what nitrogen the plant needs in July and not later than the first week in August. I have flown on sixteen pounds of nitrate nitrogen as late as August 24 but most of the time this is not good. Sometimes it will pay off.

When there is too much nitrogen in proportion to the other food, the nodes will be from six to seven inches long and a square won't set until a later date. The stalk will be large and you will have to pick it late.

It's usually better to make the crop early and get it out early but ever so often when a late frost comes along, the late cotton will make as much as the early cotton.

I think we have a good lead on what makes a five-lock boll. We know now in 1977, that by properly feeding nitrate nitrogen the late bolls of cotton that make in the top will be as large as the early bolls on the bottom of the plant in the first part of the year. This will add many pounds of seed cotton to the production.

I have found that the best production does not come about when you put the greatest amount of plant food per acre, but depends on what mixture of foods is used and when you put it out. One farmer may spend $25.00 per acre for fertilizer and make less cotton than his neighbor, who does not spend one cent. A farmer must put out the fertilizer that the plant needs when it needs it, not sooner or not later, but when it is needed by the plant. There is no set rule because each piece of land is different and needs different treatment. Just because Bill Jones's crop makes good with one hundred pounds of anhydrous ammonia and four hundred pounds of 4-11-11 is no reason at all for Sam Smith to use fertilizer of this type. Today top profit farmers use soil tests and tissue analysis to determine fertilizers to buy and at what time they should be used.

The real answer to best production is in the proportion of the food the plant receives, regardless of whether you put it in the soil out of a sack or whether it was put in the soil by nature.

With the great demand worldwide for fertilizer the farmer has a responsibility to use it wisely. He will also save money when he learns to use feritlizers at the time the plant needs it to get best results.

My notes May 22, 1977: "The laws of farming never change.

Proper feeding results in the five-lock boll.

They are not a respecter of persons. These ideas, when followed, guarantee the best that the land will produce in any year."

To improve our production requires thought and action. We have gone about as high as we can go by our present methods. We must learn more about what makes a plant act the way it does. We know so little today that the surface has hardly been scratched, as to the possibilities for increased yield new methods will bring about.

THE SEARCH FOR CORRECT NITROGEN USAGE

In 1968, after learning how to handle water so the plant would make a higher yield, I saw there was still a weak link in the production chain. The weak link was plant food and I did not know what the plant needed—the present method of soil analysis did not tell me. There I was, not much better off than before. One thing I had found out and that was that I didn't know what the plant needed. I didn't know anyone who did. I realized that we all forget from year to year. I decided to write down the ideas and happenings that occurred from day to day and maybe some good would come of this. There was no one that I could turn to who could answer the question.

I had gone through all the research material I could find that related to plant food needs. There were some writings that spelled out the needs but these writings had been overlooked and ignored. Some of the following books on research of the cotton plant were studied and their findings applied in a practical way in my work. *Fertilizer Technology and Usage* by M. H. Vickers, G. L. Bridges and L. B. Nelson, pages 88-107; *Advances in Production and Utilization of Quality Cotton Principles and Practices* by Fred C. Elliot, Marvin Hoover and Walter K. Porter, Jr., pages 190-209; *Hunger Signs in Crops* by Howard B. Sprague, pages 69, 70, 97; *Nitrogen in Crop Production* by Jim Valentine, Texas A&M College. I have studied other publications too numerous to mention.

We have for our ultimate goal production one and one-half bales of cotton per acre. We always hope for two bales of cotton, which we seldom get. This is the limit to our production thinking, because we haven't studied the daily needs of the plant and just blindly put out a food that frequently injures the production possibilities of the plant rather than aid it.

That one and one-half bale figure should be our smallest expected yield from now on. With the knowledge available in 1977, our goal should be four bales of cotton per acre and higher, when we study the plant's needs and supply the food the plant needs at the proper times.

It is a well-known fact that too much nitrogen in the plant at an early stage discourages the setting of fruit and aids the plant to grow. Many years ago I began to suspect that nitrogen might be the plant food that could be adjusted to make the plant grow as we wanted it to and make cotton produce an earlier and more bountiful crop.

Many years ago Dr. Edward Greco had asked me about my fertilizing program. I told him I put on one hundred units of nitrogen to the acre and that was it. He suggested that might not be the right thing to do. When NH_3 came out in 1949 I used it and bought a thirty-thousand-gallon tank. It worked wonders the first few years and I thought we had found the answer.

About 1950-52 I worked out a program of fertilizing my cotton using leaf tests, under Dr. N. D. Morgan's instructions, and we made some good yields on some cuts for several years. I was never satisfied because I didn't know why. I could never understand why one field would make lots of cotton and another not make much.

In my July 1963 notes I write: "Yields may be limited to a number of factors which include too much water, too little fertilizer, too much fertilizer at the wrong time." By September, I had decided that the timing was the important thing, comparing the proper feeding to the running of a car. My notes say: "Timer on a car—a new car can be purchased with a perfect motor, plenty of gas in the tank, the oil perfect, the car greased, clean, and ready to run but if that car is not timed correctly, it will not run properly. The same thing applies.

"It looks as though today (July 11, 1963) we have fertilized just right. We put out two hundred pounds of 5-20-20. This did the job well. The nitrogen was then placed deep and most of it in the middle so the cotton would not reach it until after the squares appeared. Approximately 100 units of nitrogen was put on all the cotton in various forms. I believe the lack of excess nitrogen causes the plants to begin setting fruit. It could be that it worked this way this year because it was dry; had there been lots of rain it might have grown tall. Today it is not more than waist high and loaded with fruit. Average yield about two bales per acre."

You can see how my thinking developed over the years as you go back and read my old farm notes. July 6, 1964: "The oil well cut is small and covered with squares. Joe Marselius said it might be to its advantage for it being small. Let's watch and see if Joe is right. We were watering this cut." August 18:

"Had two inches of rain. We stopped irrigating for the year." Idea: Cotton can make two or three bales cotton per acre in the same number of making days as easily as it could make one bale. The food has to be in the proper balance. What is that balance? The oil well cut described above produced thirty-five bales of cotton or two and one-half bales of cotton per acre. This cotton was planted solid. Two and one-half bales of cotton per acre of solid planted cotton equals three and eight-tenths bales of cotton of two and one cotton.

On August 20, 1965, Dr. N. D. Morgan of the American Potash Institute came up, made tissue tests, and discussed plant food balance. According to his tests I was using too much nitrogen too early. "It should not have too much nitrogen until it starts to fruit and then more cotton will be made. Having water when it is needed is essential to top production. July and August are the months when cotton will be injured by drought. Top production cannot be made when even short periods of moisture stress occur."

We got food into the cotton plant this year (Sept. 1968): "Not a perfect job, but a start. Getting food into the plant is not a haphazard job and I don't know anyone who knows how today. Some day there might be some one who will find out. I believe I could make a start if I had technical help—otherwise I won't live long enough. We picked over the first time and got 1185 bales on eight hundred acres. The white flies cost us four hundred bales of cotton."

On July 16, 1969 I wrote in my notes the feelings that many farmers must have had. "It is my opinion that we spend more money than we should on fertilizer but we know so little about the proper amount and what kind. Soil tests are good but we don't know how to read them. Tissue tests are good but we don't know how to recommend after we tissue test. We have so much to learn."

I had a wonderful crop of my own in 1969. My notes for September 14, 1969, said: "I am pretty sure I have some cotton that will make four bales per acre now." Dr. N. D. Morgan, Dr. Pinckard, and Ray Smith wanted to have a field day and show it off. I was against doing this and told them to let it ride for a year. I thought I knew what took place or caused it to set such a crop. After thinking about what happened, I realized it was an accident. I could not duplicate it another year without more knowledge about what went on inside the plant.

On October 13, 1969, Ray Smith and Dr. Morgan came up. "Dr. Morgan thought cut No. 1 on Havana was the best he had ever seen. Ray Smith wants to get the *Delta Press*. Both think I am on the right road. We can duplicate this next year on almost every cut."

Dave Pearce, Commissioner of Agriculture for Louisiana, Gordon Van Hoose, Sr. and Kay Chandler came up on October 17. On October 30, 1969 "Dr. Morgan, Dr. John Cox, J. Y. Oakes, Ted Carroll, Dr. David Jones, and David Caldwell came up to discuss plant food and it rained two inches right after they got here."

Our present methods of farming are obsolete. My January 24, 1970, notes say: "Each acre of land will produce several times what it does now at a lower cost per unit. The reaction to this new way of feeding the plant will be the largest thing agriculture has experienced in my lifetime; I've been with agriculture since 1928 and then we had a single mule, one man, and a straight stock with one sweep. Most of the knowledge is here now and in the future will be used but it is not accepted yet. This old way has played out. This old way should play out entirely in five years, for it will be a scramble for survival among the farmers by that time." In 1977, as I write this, the change is taking place and will continue. Some older farmers would not poison insects and went broke. Some farmers would not water their crops and went broke. It seems that farmers are at a crossroads again. We must increase our production to offset increased costs or we all will go broke.

In this day of large acreage it's impossible for a planter to give the close supervision to his crops that is necessary for the best production. He hasn't the time to devote to the many details that are required. He must have help. The job of looking after the details of the plant food needs so that you can make more production is just as important as looking after the mechanical needs of the tractors so they will run. This takes close supervision by people who know what food is needed.

Dr. Al Lengyel (Lengyel's Agricultural Consulting Service and Laboratory at Phoenix Arizona) advised me on the 1970 crop. I took one half of forty-acre cuts from one side of the farm to the other, following what Al Lengyel said on twenty acres and on twenty acres maintaining plantation practice. I was not to put out any nitrogen before planting. We were trying to control the nitrogen content in the plant so that the plant would shorten the joints and cause the squares to stick. This would control the excessive growth of the plant. Dr. Lengyel had some previous experience feeding the plant to correct the balance of food within the plant. He would adjust the feeding to whatever the petiole samples indicated the plant needed.

We made improvement on the 1970 crop. On August 17, Dr.

Morgan and Dr. D. Fielding Reid, president of American Potash Institute, came up and spent the afternoon. Dr. Morgan is of the opinion that the plant should stress slightly for nitrogen in June. Our present method of feeding will not make a big crop.

On August 19 I wrote: "I am convinced we do not use nitrogen correctly. When we do it right, our production will double and under the same conditions." And on September 3: "I am convinced too much nitrogen in early stage is the reason for the big stalk." Later on I wrote: "When the food is right the cotton sheds its leaves of its own accord." And on September 15: "Two hundred flags were tied on two hundred white blooms today to see if they will mature." Those blooms did make well.

In describing the Huddleston Place field in September 1970, I write: "There are some places where cotton is wonderful and places where the cotton plant is too big. We put twenty-two pounds ammonium nitrate on stalks last fall to rot them. Was in corn for four years. This spring we broadcast two hundred fifty pounds of ammonium nitrate. This was too much nitrogen for cotton. We have no idea what it will make. Some places the plant is seven feet tall." That cut was worse than terrible that year.

Our soil has so much plant food in it that the soil will have to be brought back into balance before a maximum crop can be made. This might take five years. The tall cotton was due to too much carryover of nitrogen from previous years.

For the 1971 crop, I took the same land as I had used in 1970 so that a true comparison could be made. I did not put out any fertilizer on those cuts preplant. I sent petioles each week to Al Lengyel on the cut of mine and the adjoining cut (fertilized the plantation way) that was used for comparison. I kept my eyes open as to what the plant looked like and took slides. I tried to put out enough nitrogen on cotton to keep it growing and holding squares but not enough to throw it into a vegetative state.

On August of 1971 I wrote: "Plant black land in April. Keep trips off. Keep hoppers and plant bugs off. Have ample plant food except nitrogen in soil before planting. Add nitrogen in ample time and ample amount to keep plant growing and to stick all squares. Furnish ample moisture but don't soak land." In 1977 I still think this is good advice to follow.

I continued to work with Dr. Lengyel. My notes on July 28, 1972: "I sure would like someone to help me see. Dr. Morgan is so busy and all others think I am nuts. Will write the best I can and as I see it. If we can overcome all the problems of cotton production, we will have lots of cotton that will make several bales per acre. Can we overcome them?"

By testing the petioles, Dr. Lengyel could tell which plants were in range and ready to produce and what food was needed to keep the balance within the plant more favorable for fruiting. When there is too much moisture and too much nitrogen the plant becomes vegetative with dark heavy leaves. This will show up in the petiole test. We can feed the plant sugar (which contains carbon) through the leaves. The plant will take it up to bring the food within the plant to the right carbon-nitrogen ratio and the plant will continue its normal growth development of fruiting. It will not go on making a big stalk with no fruit on it. If the ratio is not brought into balance, the plant will make nodes six to seven inches long with large leaves and long petioles. In all probability the young squares on the plant will shed and this contributes to more growth and more size of stalk.

Dr. Rouse Caffey, Louisiana's new Experiment Station Director made a trip to Arizona with me in 1972. My August 7 notes: "Returned from Phoenix. Our land is so much better than the desert, but we know so little—we just think we know a lot! Our stalk is too big, and it hasn't got the fruit on it. A plant doesn't set fruit if there is too much nitrogen available early. After it sets the fruit, then nitrogen can be added in the proper amount to hold that and then set more. This is what we have to determine, and I doubt if I ever find out because there are not enough years left for me. Today no one else seems to be interested. Dr. Caffey has an open mind. I think he wants to find out. If he does, a new era of farming will open up for those in the mid-South. Dr. Caffey is very unusual. We left the motel at 4:00 A.M. with Dr. Lengyel and we stayed in the cotton fields all day. Ate breakfast on the hood of the car at eight and lunch at noon next to irrigation well, with eight Indian children and one squaw swimming in the irrigation ditch. We met one farmer in the afternoon. It was late in the afternoon before I noticed Dr. Caffey hesitate about getting out of the car. I admitted I was given out and he said he was too, but he didn't want to admit it. The temperature was around 115°, more or less, and it seemed like more. It's most unusual for a man to want to see and investigate as much as Dr. Caffey. I think this will be a great help to the farmers."

Dr. Woody Miley from Arkansas went to Arizona to visit Dr. Lengyel in 1973. Dr. Miley said he understood the method and would introduce it to his state of Arkansas.

On August 10, 1972 I had written: "Too much nitrogen early will definitely delay fruiting. We have so much nitrogen in the ground that it is far better not to put out any at first and later to add nitrogen as it is needed. Ammonium nitrate put out over the top of cotton when it's small and before squares start setting will delay the fruiting and cause a big stalk and then make a crop in the top. Anhydrous ammonia placed in the bed will cause fruit to be delayed. It is better to put the NH_3 in the middle as deep as possible so the roots won't reach it until later. This will cut down on the size of the stalk and the bolls that stick will help hold the stalk down. NOTICE—NOTICE—NOTICE—nitrogen early will delay fruiting. Most of these ideas are old and proven but we don't follow them."

"One of the real joys of learning and progressing is to share it with others. When we learn we must write it down so it will not be lost. If we keep it to ourselves we become stagnant and there we have lost and so has posterity. If we can discover even one small truth . . ." and my search for nature's truths continued.

On September 7, 1972 I write: "It is so evident that too much nitrogen early has delayed the boll setting. This wrong application of nitrogen has been suspected by me for ten years but this year I have finally put my finger on the cause."

I have found this search for knowledge an interesting road in life. This search to find out what is really true and what is half-true as it applies to the cotton plant pays off in satisfaction and money. Most farmers are satisfied with a pittance of what is possible.

Thorough knowledge with full curiosity and willingness to learn is rare indeed. Education is certainly a lifetime process. Some of the most highly educated people have no formal education but they are willing and want to seek the highest—the truth. They ask why and will look at results. Many people are like tall willows in a breeze. They bend every way the wind blows.

How the plant reacts to feeding will introduce a new era in farming and will eventually cause a dramatic change in agriculture, just as the tractor changed our way from the single plow, one mule and a hoe to work our crops.

HOW THE PLANT REACTED

In summing up this information and trying to put it into a form for the cotton producer to get some benefit, I want the reader to know I do not have all the answers to these problems. Those who have farmed for a number of years know this statement to be true. Nobody has all the answers and they never will have.

I have tried to describe in a general way how a plant grows, what makes it grow, what makes it produce fruit, what causes the fruit to stay on the plant, what causes the fruit to be thrown off the plant, what makes it stop producing, what makes it speed up production. Why? Why? Why? What makes it act as it does? Why?

What happens when too much moisture is present in the soil? What happens when there is not enough moisture in the soil? What happens when there is too much nitrogen in the soil at a certain time in the life of the plant? What happens when there is not enough nitrogen in the soil at a certain time in the plant's life?

These problems that are being discussed are not new. They are as old as the plant itself and we don't know how old that is. They were with the plant at the beginning.

These ideas are based on only a fifty-year span that I have lived through, but I think the plant's food needs are the same as they were fifty years ago or will be fifty years from now. To supply these needs becomes more urgent each year, as our soils are depleted of plant food.

In November 1971 some of the forward-thinking men in agriculture got together in Shreveport, Louisiana and discussed this problem. All agreed the future farmer would have to have someone to advise him of the plant food needs and to look after the technical analysis of the plant. From this meeting the Southern Agriculture Consultants Inc. was organized. Those attending the first meeting were Dr. N. D. Morgan of American Potash Co., Dr. Al Lengyel of Phoenix, Arizona, W. Ray Smith Agriculture consultant, E. K. Chandler, American Plant Food Co., Galena Park, Texas; Dr. Melvin Stevens, Northwestern Universary; Harris Barnes, *Delta Farm Press;* and Dan P. Logan, Sr., cotton farmer. Dr. Hal Lewis, professor of biochemistry at the University of Tennessee could not be present. It rained that fall and the next four falls at such regular intervals that we could not get any records worth recording. We knew how to grow the cotton but because of the rains we could not harvest it.

On September 14, 1973, a group was here at my farm at Gilliam to see the progress made in plant feeding and a paper was typed to explain what the object of my work was:

"If there is not a boll on every joint, it's my fault and not the

Leading cotton experts from Texas, Arkansas, Mississippi and Louisiana, inspecting cotton field at the Dan Logan plantation.

plant's fault. The plant put it there but we farmers haven't learned how to hold it. The food being out of balance is the reason it's not there.

"In every field the bolls are set too far apart and too high up on the stalk. The reason is food imbalance. In no case is the number of bolls over half as many as there should be. If the food and insect control plus moisture had been right there would be a boll at every joint.

"Moisture has not been a limiting factor except in isolated cases. After learning how to handle water several years ago, very few times has moisture been the limiting factor.

"The plant responds exactly the same way to food and water every time. It's just as practical for us to take a tissue test as it is for a medical doctor to take a blood test." Those attending the meeting: Dr. Tom Burch, Dr. Joe Sedberry, Jr., Ted Carroll, Dr. Rouse Caffey, Dr. John Roussell, J. Y. Oakes, Dr. N. D. Morgan, David Caldwell, O. S. McFatter, Dr. Melvin Stevens, Richard Maples, Dr. Woody Miley, Roy L. Waddle, J. Chester Stevens, Andrew Whisenhunt, J. S. Stone, Robert Pittman and Harris Barnes. Some of the guests were from Louisiana State U., Northwestern and the University of Arkansas.

If enough food is available in the right proportion the plant will set more fruit faster and won't go to too much stalk. I think too much nitrogen in relation to phosphorus, potash, secondary and minor elements is the cause of extra growth and not setting of fruit. Poor insect control can also be a cause of poor setting of fruit and excessive vegetative growth.

On August 6, 1974: "There is no question about this method of plant feeding being correct. The growth of the plant can be controlled and the plant will react the same way every single

Group of experts discuss findings at meeting on the Dan Logan plantation.

time to food and water. Every single square can be set if we can do our part correctly feeding, watering and controlling insects."

September 15-20 of 1974 were spent looking at the cotton in East Louisiana, the Mississippi Delta and all over Arkansas. Some of the farmers visited were Duke Shakelford at Bonita, Keener Howard and Jack Hamilton in Lake Providence. Early Ewing in Scott, Mississippi and Dozer Lester and Bill Lester in Inverness, Mississippi, Bill Davis at 4/5 Plantation in Greenwood, Race Track Plantation in Greenwood then on up the country to Ross Ingram in Water Valley and George Shepherd at Lake Comerant. Then over to Odis Chapman at Scott, Arkansas and up for a visit with Dr. Hal Lewis in Dell, Arkansas and the farmers in that large area.

Most all these farmers used plenty of nitrogen and most all their stalks were too big. We are all about in the same boat, our plant food is out of proportion in some way. We all have so much to learn. I am pretty well convinced that nitrogen out of proportion to the other plant food is where most of our trouble lays. We just must learn to make cotton earlier and have it mature faster and cut down on the size of the stalk if we want to increase our production. The nodes must be three inches long instead of seven inches long. This can be done by feeding the plant the proper food at the proper time. We all agree with this statement but the question is how do we do that.

God put us all here to work to the best of our ability. It looks like my progress in learning how to do better is slow, but maybe if I keep looking something will be learned eventually.

On August 14, 1975, Dr. Jack Jones and David Caldwell

came up from L.S.U. Dr. Jones is the cotton breeder and David Caldwell is his assistant. After all-day discussions I wrote out my beliefs and gave them a copy. Dr. Jones gave me a book on the cotton plant that he had just compiled. The points of my observations were: The plant will make more if it never stresses. Moisture in the soil will not make the plant grow too big if the food is in balance and if insects are under control. It is possible for a plant to hold every square it puts on and bring it to maturity. We will have to know more than we do now as to what to feed the plant for it only throws off now if there is not enough food to sustain the squares it sheds. Watering the later part of August will not make the stalk grow too big as long as the food is in balance and the plant has never been allowed to stop growing for the want of moisture. This excess growth is called second growth and occurs after the plant has stopped growing and then gets additional moisture and is trying to grow and set another crop. If it has never stopped during the first growth the second growth will not appear.

I think today, August 14, 1974, that it is impractical to have enough plant food in the ground at the beginning of a crop to bring to maturity all the squares that a plant puts on. Maybe someone will find a method to accomplish this and I think they will. Today I haven't heard of anyone who is working on it.

It's impractical to feed a calf enough food in April or May, with just a small amount six weeks later, to do him through August and September. The plant needs the proper food in balance at the proper time in order for it to do its best job of producing. If a plant can get this food it will bring every square to maturity. The plant needs daily attention to its diet.

When we test these plants once each week we will almost know what the plant needs, or at least we won't be far wrong. We will know a whole lot more than when we guess at it.

Do not let the plant ever stop growing and setting fruit. Ample moisture will not cause the stalk to grow too large if the food is in balance and the bugs are kept off. We must be concerned with the food needs of our plants if we want to produce more per acre. We just must find this out and do it whether we like to or not.

It has always been hard for some of us to change our method of working whether it pays us to or not. We just naturally don't want to change our past ways.

A story brings out this not wanting to change. It was thought best that Carol, one of the hands on the place, should move over to the other side of the farm where he would be close to the pasture where the stock was kept. Carol did an excellent job of running the gasoline motor that kept water for all the stock and he was very dependable. Carol did not want to move. He lived close to a bayou that had lots of fish in it and he kept several hooks set to catch big catfish and loggy head turtles. He liked to be near his friends and I was trying to persuade him to make the move. I heard what he was saying. "Mr. Dan is wasting his time trying to get me to move over there. I'm not going to go. Mr. Dan is just wearing out his mouth trying to talk me into that!" When farmers don't like to change, no amount of argument will work regardless of how much it will profit us. We just like to set our hooks in the same bayou we have been setting them in all the time.

On September 30, 1975, a number of visitors came here to discuss the possible production breakthrough. Farmers were becoming more interested in petiole analysis and keeping up with the food in the plant. Some of those present were Dr. Gene Woodall, Eugene Sutton, Ron Chapman, Otis Chapman, Dr. Woody Miley and Doug Chapman from Arkansas. From Louisiana Dr. H. House Caffey, Dr. Olin Curtis, Sherman Phillips, Dr. N. D. Morgan, Dr. John Roussel, David Melville, Fred Andries and J. Y. Oakes, Danny Logan, David Caldwell and from Texas, W. Ray Smith, agriculture consultant from Midland, Texas, and David Robinson.

By the end of 1975 considerable progress had been made toward shortening the joints and causing the squares to stick. We had found that the need for nitrogen in the plant is high, for this is the time so many squares are setting and the bloom is heavy and all the bolls need nitrogen to fill out to maturity. We knew what food was in the plant by our weekly petiole analysis and we tried to keep the right food within the plant for the best fruiting. When the petiole sample showed that we were getting high with nitrogen, we were able to counteract this by foliar feeding sugar. We did not do a perfect job but progress was made, enough to show us we were on the right road.

More and more we realize that all these things work together —timing of feeding, trace elements, moisture. Just the absence of one nutrient alone with all others in balance will offset the growth and development of the plant. According to Dr. Morgan the timing of the feeding of nitrogen will determine whether the crop will mature early or late.

In 1976, Dr. Dysart E. Holcomb of the Pennzoil Corp. came up and wanted to work with me on soybeans and cotton. They believed that farmers were limited in their production of these crops and they wondered what they could do to help break this production barrier. The Pennzoil Corp. owns so many of the mines over the face of the earth where the basic plant foods come from. I explained to Dr. Holcomb that I was trying to find out why this production could not be made to rise and felt that we had run across ideas that looked promising but we had a long way to go.

Dr. Lengyel and I made a trip in 1976 over the mid-South, introducing this method to interested farmers. We started in Shreveport then on to Lake Providence and Sicily Island, Louisiana, to Greenville, Mississippi and Delta Pine Land in Scott, Mississippi. We talked to interested Clarksdale farmers with John Mc Kee at Friers Point, and interested farmers around Greenwood with Billy Harris, then wound up the week with a big meeting of the Arkansas farmers at Little Rock with Dr. Woody Miley and Otis Chapman. We made some progress with opening the eyes of some over the mid-South.

My notes: "On May 19, 1976, the temperature was 52° and it has been this way all over the mid-South. I doubt if the mid-South can make a crop this year for so many of the farmers haven't got a stand yet. Our cotton is growing but we are in the southern part of the mid-South and are just a little warmer than most."

Dr. Lengyel's May 1976 report showed that our plant food was more in balance now than it had ever been. Maybe we are making progress.

July 14, 1976: "The plant is getting ready to really perform in all its grandeur. I can't help contrasting this crop with one just thirty years ago in 1946." On July 14, 1946 I finished chopping cotton and there was not a bloom on the whole farm. The boll weevils had everything. I was the overseer but the boss told me to do what I wanted to as he didn't know and he was leaving it in my hands. It looked too bad for him to make any decision.

I got a plane and enough Calcium Arsenate to poison everything seven times on a five-day schedule. When the showers washed the poison off we put it back on at once. We were considered nuts and everyone knew we both were crazy but we made a bale of cotton to the acre. Our neighbors didn't poison and it took from ten to fifteen acres to make a bale of cotton.

On September 29, 1976, Dr. Woody Miley of the University of Arkansas came down here and met Dr. Rouse Caffey of L. S. U. and his staff, Dr. Ken Tipton, Dr. John Roussel, David Melville and David Caldwell. Dr. Miley made the comment that putting two bolls of cotton where one grew before seemed to be very close and he paid me the compliment that a lot of this progress was due to my persistence. I, too, think very strongly that the light of progress is showing through the production barrier, due to following this method.

To illustrate the point I am trying to bring out: When I was a boy I got a job cleaning out a well. I would fill up a bucket with the dirt in the well and it would be drawn up with a windlass and emptied on top. Each bucket of mud would put me deeper in the ground. When I looked up I could see the stars in the sky, even though it was daylight. After working at this job, I think strongly that I can see the stars of hope even at this early stage of the breakthrough for increased production for the cotton farmer. More and more farmers all over the cotton belt are beginning to get a glimpse of the tremendous possibility for high yields.

On October 23 and 24, 1976 I met with Dr. Al Lengyel and E. K. Chandler in Phoenix, Arizona to talk about the future of the plant analysis method and what procedure to follow. In my opinion, it will take ten years for farmers to begin to put this work to practical use.

Arkansas is making greater strides forward than any other state, due to the work of Dr. Miley and Dr. Maples and Dr. Woodall. They are trying to train people to carry on the work and help with it.

Louisiana State University is doing some research on the needs of the plant as indicated by petiole analysis. They are beginning to look into this.

Missouri initiated a cotton monitoring program in 1977 in which one hundred farmers participated.

In 1977 Tolly Burns, who retired from teaching agriculture, conducted work on moisture to tell when the cotton plant needed water and when to irrigate, if you have water. This work proved valuable to me for it was the first time that I could tell when to start irrigating my crop and just where the moisture level was in the ground and then I could predict just about when the crop would burn up or start throwing off bolls and squares due to moisture shortages. The details of this study were outlined elsewhere in this book.

FROM THEN UNTIL NOW

In 1915 my mother took in boarders and gave people meals when they came to the Court House in Benton, Louisiana. There were no restaurants. Traveling was done on horseback or by buggy. Our house set on a city block with a fence around it. The horses were turned loose in our yard to graze and wallow in the sand while the people attended to their business in town.

Our house was the general meeting place for the whole vicinity. There were two stores in town that sold all the needs anyone might have, from safety pins to coffins. However, most coffins were made by friends at home. That was one of the things the neighbors could do was to make and dress the coffin after one died.

A typical farmer would ride into town on his horse. Behind the saddle a rolled-up riding slicker was tied. In case of rain, this slicker was large enough to cover the rider, the saddle and hang below his feet that were in the stirrups. You could not use this slicker while walking for it would drag the ground. A scabbard with his pistol hung from the saddle horn where it was readily available. If he left the horse he would tie the slicker over the saddle and put the pistol in his pocket. I never heard of anyone stealing a scabbard or slicker or saddle.

When automobiles first came into use, almost every farmer, when he came to the gin with his cotton, carried his shotgun. If by chance he met a car that would certainly scare his mules or horses, the shotgun was a persuader for the car operator to pull out of the road with that loud popping thing until the farmer got by.

There wasn't much difference in the way people farmed in 1915, or 1920, or 1928 than the way they farmed fifty years before. All we did was cut the stalks if we had a stalk cutter or drag them down with a long pole or log. We would run a middle breaker in between the rows that threw dirt both ways or we would take a turning plow pulled by one or two mules and put two furrows on each side of the row throwing the dirt to the middle. This procedure was called listing. This would leave a piece of land where the old stalks stood that was called a baulk. A middle splitter was then used pulled by three or four mules. This middle buster would throw this baulk out on top of the land that was previously thrown up by a turning plow. This would make a high row.

The land would be ready to plant then. We would then pull a log over the top of two rows at a time with a mule on each end of the log. This would leave a fairly smooth place for a mule to walk and pull a Georgia stock with a fourteen-inch sweep that would knock all the trash off the top of the row and then the one-row planter, pulled by one mule, would have a clean place to plant the seed.

When the cotton came up we would take one mule, hook him to a turning plow and go down the side of the row and come back on the other side of the same row. This narrow ridge of land left would be where the cotton was planted. We would go through the cotton then and hoe out all the weeds and grass and thin the crop the way we thought it should be thinned. A Georgia stock was again used to run the middle out and then a double shovel was used that plowed one-half a row each time and again left the cotton on a high row. We would try to plow through the cotton once each week until picking was started sometime in the month of August. We would stop plowing and go to picking.

This would mean we would make fifteen to twenty trips over each row to break, plant, and cultivate a crop. I doubt if we would have the patience and time to make all those trips to complete a crop now but we didn't have a television program to look at at a certain hour. I am not saying whether it was better then or better now. I am just saying how it was done then. You be the judge as to which way you would like it best. People didn't need to take sleeping pills so they could sleep then. All they needed was to get to a table where there was plenty of peas, buttermilk, a liberal slice of home cured ham topped off with a big pone of corn bread with a pitcher of syrup. After this was finished sleep just came naturally, and with just as sweet dreams as if Guy Lombardo was playing near by.

Every man had his piece of ground to work, with his house in the middle of it. He broke this land, planted this land, cultivated and gathered the crop with the help of his family. Depending on the size of his family he planted from ten to twenty acres of cotton, from five to ten acres of corn and his own garden of whatever he wished—cushaws, pumpkins, melons, sweet potatoes,

Irish potatoes, tomatoes, collards, okra—and raised his own sugar cane for molasses. The owner of the land got half the cotton and half the corn. The other crops belonged to the farmer. The owner furnished the land, tools, house, seed, and furnished the farmer the necessities including a doctor if necessary. For work done other than on his crop, the farmer got paid in cash.

The advance made against a crop was paid in Brozine (plantation money). The farmer could buy whatever he wanted, when he wanted it.

Not all the houses had a road to them. There was a path or a wagon could be driven over the ends of the rows along the ditch banks. The farmers walked to the Commissary or to the store. No cars were owned by me or the farmers at that time. Many of the workers owned horses that they rode to church and town. Some had buggies and wagons. They all prized a horse that could single foot.

They all had a single-barrel shot gun. On rainy days they hunted anything they could find. They all bought New Club or New Chief black powder shells that sold for forty cents a box of twenty-five shells. When these shells were shot the black smoke would roll out the end of the barrel so thick and heavy that you had to move to one side real fast to see if you had hit what you shot at. These black powder shells would kick so hard that the gun would be hard to hold.

The bayous and canals and lakes were full of fish and the workers always had some hooks set out to catch a big catfish or a big logger head turtle. All the young boys would have rabbit traps to catch rabbits or 'possums.

This was the life on a cotton farm in 1928, just as it was

fifty years before in 1878. The depression came and caused a change and this era passed on and never will be back.

Some of the finest Christians I have ever known are some of these fine colored people and lots of their descendants are here now.

I had been on the farm one year when the Great Depression started. The owner of the plantation mortgaged his plantation in order to feed these hands of his. None were made to leave and no one ever got hungry. No one had any money. Cotton was four and five cents per pound but we had plenty of hogs to butcher, plenty of beef to kill and we raised the other food.

One thing we always used as a backlog. We always tried to keep a crib full of corn. Mr. John Sentell, one of the fine early men in this section, used to tell me there never was a poor man if he had a crib of corn. If he had a crib of corn he had plenty of bread. He could fatten his hogs and have plenty of hams and bacon and lard. He could fatten a calf and have the best steaks and roast. He could feed his chickens and have plenty of fried chicken and plenty of eggs and fatten all his turkeys and geese and feed his mules so he could make another crop and feed his horses so he could go to church on Sunday. He would say, "Dan, there never was a poor man who had a crib full of corn."

I wonder now what would happen to all of us if another depression would appear like the one in the nineteen thirties. Could we survive? It's a possibility we have lost something since that time.

Then we didn't have any money but we had something to eat. Maybe today we would have plenty of money but wouldn't have anything to eat. I just wonder which would be best.

CONCLUSION

This is a story about the breakthrough for getting a cotton stalk to set two bolls where one was before. It raises many questions for your consideration: How a plant grows, what makes it grow; what makes it produce, what makes it hold fruits or throw off or cut out. What causes it to speed up reproduction? Why does it act as it does.

We don't have the answers to all these questions and probably won't for years to come. But we can say now that the plant analysis method of growing cotton will work. If I am not mistaken, the system will work every time it is used and it will work anywhere it is used. This method goes down to the basic makeup of the plant itself and the plant reacts exactly the same way each time to a given condition.

There are several powerful factors in cotton production. One is weather, which is external to the plant. Another is heredity which resides within the plant. And the one which we are concerned with here is food that causes the plant to reproduce to the fullest. My notes for the last twelve years tell the same story. The stalk is growing too big and is growing too fast and not setting the fruit as it should. We blame this on insects and the weather. The real cause is food imbalance, so therefore the plant could not set fruit.

The plant breeders have given the farmer a plant that will produce well. The chemist has done an excellent job making available chemicals that will control grass and weeds and bugs. Keeping the weeds out and bugs off and furnishing the proper amount of moisture is a mechanical process that the farmer must see after. The proper plant feeding as indicated by the petiole analysis will cause the other factors to pay off in increased production and more profit on the crop.

I really believe the plant never changes. I believe that the plant's need for food has not changed since the beginning of time and that the plant responds to the sun, water and elements as it did 10,000 or 15,000 years ago. If this is so, we are not faced with a new problem of increased production but an old problem that has existed from the beginning—that of learning nature's truths.

Nitrogen, potassium, phosphorus, secondary and minor elements go to make up all plants. These elements traveling to the leaves of cotton come in contact with the sun's rays and cotton fiber and cotton seed are produced. These same elements going to the leaves of the corn or wheat or barley or rice or tomatoes produce fruit after its kind—the same sun and the same food but different seed. This principle applies to all plants.

The use of the petiole analysis has taken the guesswork out of when and what to feed the plant. It's just that simple.

The farming ideas I am writing down cover only a fifty-year span of my life. The problems I had were the ones that were current one hundred or two hundred years ago. Great progress has been made in the last hundred years, and much of that in my lifetime.

Men have to be free to think and dream and plan if they are to move forward. The last few years (on the scale of time) is the first time in history that man had been free to progress. Man had been held down. He was at a low state, as he had always been until Jesus came to teach men freedom. Man had never been free spiritually or socially, nor had he been free politically or physically. Jesus taught him how to be free spiritually "And you will know the truth and the truth will make you free." This upset the established way of life and Jesus was put on the cross for his thinking.

Martin Luther and his associates put the teachings of Jesus into the hands of the common man. Benjamin Franklin and his associates snatched the scepter from the hands of the tyrants and kings. They were instrumental in getting the Declaration of Independence signed and the Constitution ratified, to insure freedom for the common man politically.

From these three major moves, man has come to the point where he is free to think and dream and plan as God intended him to do from the beginning. Without these events it would have been impossible for new thoughts to take root and it would be impossible for us to have radio, electric lights, or a computer or to put a man on the moon or to increase the production of our crops. We cannot name all the marvelous things that are taking place now. There will be many more to come. Why? Because men are free. These things could not be accomplished in any dominated country.

We will only make progress and find out additional truths if the people as a whole are left alone, perfectly free to pursue whatever line of work they wish or feel that they are capable of doing.

EPILOGUE

The following radio talk, "How Farmers Lived Seventy Years Ago," was broadcast on KRMD in Shreveport, Louisiana, on Sunday, March 5, 1978. It is one of a series of such talks, the main purpose of which is to give our young people a pride in their heritage—and to inspire them to find better and more efficient ways to feed the world's ever-expanding population. This particular talk describes just how the food chain worked in the old days—and how it will have to work in the future if the farmer is to survive and prosper:

Today there is not much difference in the way people live in the country and the way people live in town.

We have electricity, water, gas or Butane. We have asphalt roads, T.V., telephone, radio, Citizen's Band radio in cars, homes and on tractors.

We get our food from the chain stores and supermarkets just like the town people do. We don't work gardens, or milk cows or kill hogs and make lard and ham and bacon any more. Neither do we go to the mill and have our grain ground into meal or flour to make our bread. Instead of milking the cows, we get milk in a gallon plastic jug.

This practice of the farmer buying his food from the store began about fifty years ago. We farmers have succumbed to the course of least effort. I am not sure this is the best way, but it is the easiest way.

Everybody in the United States has more every day than King George did in England at the time of the American Revolution. We have all the good material things of life. We want for nothing. Life is so easy now.

Sixty or seventy years ago and on back, there were no gravel roads. If you traveled in the winter you went horseback or walked or went in a wagon with horses or mules pulling it. In the winter an ax was carried in the wagon to chop the mud from between the spokes so that the wheels could roll. Mules could hardly pull an empty wagon through the mud and to pull a load was out of the question.

There was no electricity, no natural gas or Butane, no refrigeration or air conditioning. In the summer, if you got a chunk of ice you were fortunate. Ice came packed in sawdust in a croaker sack.

We did not have all these conveniences we have now. Since we didn't have them, and had never heard of them, these conveniences were not missed. We were probably happier then than we are now.

We have so much now that we are unhappy about everything. We do not appreciate the many blessings that we have. Oliver Goldsmith expresses this thought well in his *Deserted Village:*

For them light labor spread her wholesome store
Just gave what life required but nothing more
Their best companions innocence and health
And her most valuable riches ignorance of wealth.

But time has altered this way of life and it is probable that we have lost a valuable asset—self-reliance.

On the farm we got up at the first bell, which was rung about thirty minutes before sun up; when the sun rose we went to work. The bell was on a stand in the yard close to the house, with a rope to pull the bell and make it ring on the porch. The second bell was rung about sunup for everybody to go to work.

Breakfast was cooked on a big black wood stove with six eyes on top. The breakfast consisted of grits and gravy, all the ham you could eat, all the bacon you wanted, eggs fried or scrambled, biscuits with plenty of preserves and jelly of any kind, a pitcher of ribbon cane syrup and pounds of butter to put on your biscuits or waffles and all the milk you wanted. Today such a breakfast would cost so much we just couldn't afford it. In those days this breakfast didn't cost one cent. We grew, processed and saved the food ourselves.

We ground our own grits and meal for bread from our own corn. We killed our hogs and had ham and bacon and sausage hanging in the smokehouse. The eggs came from the chickens that were everywhere. The preserves and jellies were put up in season. The syrup was made from our own sugar cane. The milk and butter came from our cows and when beef was needed a calf was killed.

After breakfast we went out to do our daily jobs. Milking was done every morning and night, rain or shine, in heat or cold, Sunday morning and Sunday night. The milk was strained up in big pans and crocks and set in the cooler or screened dairy. Then the utensils used had to be cleaned and made ready for the next milking. The cream had to be skimmed off the old milk and placed in crocks ready for churning to butter. The sour milk and clabber had to be fed to the hogs, chickens, and dogs.

Wood had to be cut and brought in to burn in the fireplace. The pine wood had to be split for the cook stove and brought in to fill the big wood box. Kindling wood had to be split up in small pieces to start the fires.

The girls had to clean the lamp chimneys and fill each lamp with coal oil and clean the house. They also had to prepare the food for the following meal. They had to get the sweet potatoes from the bank and the Irish potatoes from under the house. They had to get enough flour from the barrel and enough meal and get the necessary hams, bacon, and sausage from the smokehouse. If chicken was to be served for dinner, they had to kill and dress the chickens and have them cooked.

The stock had to be looked after, to bridle and harness and curry the mules and see that they were properly fed. If necessary a few ears of corn would be thrown to the hogs.

If plowing needed to be done the boys had to do it. Every so often something was needed from the store. If a grown person went to the store, a saddle would be put on the horse. If a boy went, he would probably ride bareback. Items from the store were few. Mama would need a nickel spool of No. 60 black thread and a spool of No. 40 white. Sister would want a yard of red ribbon to make a bow for her hair. A pound of green coffee was needed and a bottle of vanilla flavoring. Old Jake, who worked around the house, would want a plug of Brown Mule chewing tobacco and little Jane would always look for a Jaw Breaker or a stick of peppermint candy.

It was always the children's job to gather the eggs and find the nests that the hens would lay in outside the chicken house. Children always enjoyed crawling over the hay and finding the nests in the hay loft. Sometimes the nest would be in the wagon shed or the buggy house or under the floor of the grist mill house. A hen will always cackle after she lays an egg and her nest is easy to locate. She advertises her eggs. A guinea hen will slip off and lay in a secluded place and slip away from her nest and not make a sound. She will have to be watched real close to find her nest. She does not do any advertising and we don't eat her eggs, yet they are as good as hen's eggs. We all eat the hen's eggs every day because of the constant advertising she carries on. The children all knew you had to use a spoon to take the guinea eggs out of the nest, for if you put your hand in the nest the guinea wouldn't lay there any more but would find another spot.

Hogs were killed and the meat cured and smoked in the smokehouse. A hickory wood smoke was kept going until the meat was thoroughly cured. Lard was rendered and stored in fifty-pound cans in the smokehouse. Flour was bought by the barrel and used on special occasions and when special company or the preacher would eat with us on Sunday. It was a Sunday treat. There was meal to cook with every day.

Sweet potatoes were stored in a bank. After curing, sweet potatoes were stacked carefully by hand on a foot-thick layer of hay on a well-drained spot near the house. More hay covered this mound of potatoes and corn stalks were placed all around this and tied at the top like a teepee. Soil was then banked around the cornstalks. The sweet potatoes kept perfectly all the year even in snow and freezing weather. Irish potatoes were kept spread out on the cool dry dirt under the house. Cushaws and pumpkins were kept from freezing by covering them with the ears of corn in the crib.

It was a *necessity* to have plenty of corn. There could be a shortage of other crops and we could get along, but if there was a shortage of corn disaster was close by. If cotton failed we wouldn't have any money but if the corn crop failed we were in trouble.

My great grandmother, who moved here in 1850 with her whole family and household, used to tell me that their most important crop was corn. Her husband died in 1866 and she ran the farm and raised and educated her large family. She lived to be ninety-eight years old. She said her first job was to see that they all had enough to eat. There were no stores to buy from and their survival depended on their ability to manage and look after themselves. I wonder if we depend on ourselves like that today? The question today is, if our distribution system fails what will we do?

We were taught to raise and store food for later on, but today nobody saves anything. We throw it away. Even we farmers don't raise and store food anymore.

Mr. John Sentel, one of the great men of our area, used to tell me that there was never a poor man with a crib full of corn.

If he had corn, he had plenty of bread for his family and loved ones. He could feed his hogs and have plenty of ham and bacon and lard and sausage. He could feed his cows and have plenty of milk and butter and a fat calf to butcher for beef. He could feed his chickens and have plenty of chicken to eat and plenty of eggs. He could feed his mules so he could make another crop and feed his horses so he could go to church on Sunday.

I wonder how many would like to go back to life as described here? What would we do if we had to depend upon ourselves again—if we had to feed our families and loved ones?

The roads were so bad we couldn't get so far from home that we couldn't get back before milking time. We didn't have any money but we had plenty to eat and since we were ignorant of money we were happy without it—more so than people today with plenty of money and in the near near future, nothing to eat.

For pleasure we went to church on Sunday. The church didn't look like our big churches do today but to our eyes they were warm and comfortable, for love and friendship was there. Robert Burns explains these small churches in his *Cotter's Saturday Night:*

Compared to this how poor religion's pride
In all the pomp of method and of art
When men display to congregation's wide devotions
Every grace except the heart.

Our days were full of work when work time came, but there was pleasure also and time to play and dream and think about our future—how we could improve at our job and be more attractive so the girl down the road would like us better.

When the fish would bite the whole community would go and fry the fish on the river bank. Some would stay there fishing for several days. There was no invitation given. We just all went.

We would spend some time hunting squirrels, or, at night, 'coons. Shot gun shells were 40 cents for a box of 25 but we didn't waste them. What we shot at had to be still and we got a bead on him.

The young boys would look forward to meeting other boys and running a foot race. By the time most boys were nine or ten they had acquired a horse of some sort and there never was a boy who wasn't looking for a race, and they would brag about how good their dog could tree squirrels and how he could hold the biggest hog and how he could whip any 'coon until the 'coon got into the water.

Baseball was a game farm boys played and enjoyed.

If times like these would come back, or the days of the depression in the nineteen thirties, what would we do? Then we didn't have any money, but we had plenty to eat. Today, if disaster comes we will have plenty of money but no food.

Think about this.

I don't know where we are headed but I am pointing out where we came from; we can't look to the future unless we look back from whence we came.

It has been a pleasure to talk to you today. I am Dan P. Logan, Sr., Gilliam, La., and I have farmed the same land for fifty years. Sometimes we have prospered and other times it's been hard, but living has been a pleasure all my life. There never is a dull day for nature has so much to teach us if we will listen.